THE RIGHT DARWIN?

ALSO BY CARSON HOLLOWAY

All Shook Up
Music, Passion, and Politics

THE RIGHT DARWIN?

*Evolution, Religion,
and the Future of Democracy*

CARSON HOLLOWAY

SPENCE PUBLISHING COMPANY • DALLAS
2006

For Shari, my wife

Published in the United States by
Spence Publishing Company
111 Cole Street
Dallas, Texas 75207

Library of Congress Control Number: 2005934710
ISBN 1-890626-61-9
978-1-890626-61-7

Printed in the United States of America

Contents

Acknowledgements   vii

Preface   x

1   Darwinian Conservatism   3

2   Decent Materialism   17

3   Ennobling Democracy   41

4   Moral Universalism   71

5   Darwininsm and the Family   99

6   The Demise of Darwinian Morality   133

7   The Abolition of Man   148

8   Beyond Darwinism   178

Notes   191

Works Cited   203

Index   206

Acknowledgments

T HIS BOOK COULD NOT HAVE BEEN WRITTEN without the support of a number of people and institutions to whom I owe many thanks.

First, I am grateful to Thomas Spence and Mitchell Muncy of Spence publishing for once again electing to publish my work. Collaborating with them continues to be a pleasure. I also wish to express my thanks to the Earhart Foundation and the Olin Foundation, whose generous financial support allowed me to work exclusively on this project during the summer of 2001, the 2001-02 academic year, and the summer of 2002. Without their help this book could not have been completed by now, and might never have been undertaken. Also, I am grateful to the National Association of Scholars, which administered my Olin Fellowship, and particularly to the NAS's Brad Wilson, who helped me navigate the paperwork and who kindly encouraged my pursuit of the topic explored in this book. I am, of course, also indebted first to Concord College and now to the University of Nebraska at Omaha for providing me with an institutional home while I worked on the manuscript, and generally for allowing me to think and teach about the issues in political theory that are addressed in the following pages.

I am grateful for the encouragement provided by all my colleagues in political science at UNO and in the division of social sciences at Concord. I am especially indebted, however, to the following three Concord faculty. Joe Manzo, who was then the chair of social sciences, helped me make the arrangements to accept the Olin Fellowship and generally supported my scholarly endeavors; and John Stack, now of Providence Academy, and Elizabeth Schiltz, now of the College of Wooster, took time for hallway and lunchtime conversations that helped me hone my thoughts on the questions addressed in this book.

In addition, a number of my teachers and mentors have generously encouraged me in the pursuit of my studies and written letters in support of my applications for research grants: Gary Glenn, who nominated me for the Olin Fellowship, Tom Lindsay, Morton Frisch, Larry Arnhart, and Darrell Dobbs.

I would also like to thank my mother and father, and my mother- and father-in-law, for helping me in so many ways that contributed importantly, if indirectly, to my work. In addition, I am grateful to my wife, Shari, to whom this book is dedicated, and to my daughters, Maria, Anna, Elizabeth, and Catherine, for allowing me occasionally to withdraw from the family for intense periods of work on the manuscript.

Finally, I owe a debt of gratitude to the proponents of the new Darwinian approach to political theory: Larry Arnhart, James Q. Wilson, Francis Fukuyama, Roger Masters, and Robert McShea. While I am critical of their arguments in the pages that follow, I believe that I have learned, and that we all can learn, many useful things from their work. The debt is deepest in the case of Larry Arnhart, who has encouraged my research despite our disagreements and whose graduate seminar "Evolution and Political Theory" first started me thinking about the issues explored in this book. Surely a student can pay no greater compliment to a teacher than to reflect on his class for ten years after it has ended.

Any merits this book may possess could not have been achieved without the help of the aforementioned people and institutions. Any defects are my own.

Preface

C AN MODERN SOCIETY dispense with religion as a source of
moral instruction?

This question has arisen in every generation since the emergence of modern societies, which, by securing freedom of thought and expression, have allowed intellectuals to question the once largely unchallenged assumption of the wholesomeness of widespread religious belief. This question is forced upon us anew in our own generation, however, by the contemptuous public dismissal of religion by certain prominent scientists and popularizers of science, particularly those working in fields informed by Darwinian biology. Consider the cases of Steven Pinker, Richard Dawkins, and E.O. Wilson.

Each has achieved an enviable prominence in his field: Pinker, a psychologist, holds a professorship at MIT; Dawkins, a zoologist, is the Professor of the Public Understanding of Science at Oxford; Wilson, an entomologist, is a research professor at Harvard. Each has also won public acclaim as a popularizer of the evolutionary account of nature and man: all have authored multiple best sellers, from Pinker's *How the Mind Works* and *The Blank Slate*, to Dawkins's

The Selfish Gene and *The Blind Watchmaker*, to Wilson's *Consilience* and *Naturalist*. And each has publicly dismissed religion as unworthy of belief.

At the outset of *The Blank Slate*, Pinker suggests that the Judeo-Christian understanding of human nature is based upon a biblical account of creation that a "scientifically literate person" can no longer believe.[1] In an interview with *Reason* magazine he claims that the belief in "an immaterial soul that is the locus of free will and choice and which can't be reduced to a function of the brain"—a belief he associates with the "religious and cultural right"—is a "myth," in the sense of being "wrong" or false. "Neuroscience," he claims, "is showing that all aspects of mental life—every emotion, every thought pattern, every memory—can be tied to the physiological activity or structure of the brain."[2] In an online interview charmingly entitled "Religion: For Dummies," Dawkins contends that "you won't find any intelligent person who feels the need for the supernatural," unless he was "brought up that way." When asked whether he can see any positive contribution religion has made to the world, he replies: "I really don't think I can think of anything; I really can't." For him, religion is a "waste of time" to be classed with "astrology, crystal-gazing, fortune-telling, things like that."[3] Wilson, while he admits to getting a quasi-religious satisfaction from the scientific exploration of nature, similarly views traditional religion as a "waste of time," dismissing the biblical account of human origins as a "scribe's myopic view of the subject written 500 years before the birth of Christ!"[4]

Indeed, for these prominent scientists and public intellectuals, religion is not only false but harmful. Pinker and Dawkins in particular, choosing their illustrations from the most spectacular of recent evil acts, denigrate religion as a source of hatred and violence. Responding to the fear that morality will be undermined by the belief that it is merely a "product of the brain," Pinker argues that "[i]n practice" such materialistic views are "less dangerous than the idea that morality is ultimately vested in the commands of a religious authority. 9/11 is only the most recent example of a case where morality derived from religion leads to horrible atrocities."[5] Dawkins likewise notes darkly that the "utterly disgusting" story of "Abraham almost sacrificing Isaac" is "iconic" for Judaism, Christianity, and Islam. Without religion, he says, we would have "paradise on earth," for "there would be a much better chance of no more war. Obviously nothing like 9/11, because that's clearly motivated by religion. There would be less hatred, because a lot of the hatred in the world is sectarian hatred."[6] On this topic Wilson is characteristically more temperate. While he notes that God had "plenty" to say to Moses about "tribal loyalty and conquest," he concedes the association of "charity, tolerance, and generosity" with "traditional Christian teaching." He also suggests, however, that biblical religion is inferior to modern science in fostering respect for nature, one of his primary ethical concerns.[7]

As such lines of criticism suggest, these are earnest men—not at all like the cynical, amoral atheist of popular imagination who, having seen through religion, claims to have discovered the groundlessness of all human values. They may not believe in God, but they do emphatically believe in morality, or at least a certain morality. Pinker endorses a "humane" respect for the "dignity" of those who suffer that shows itself in seeking cures for their ailments.[8] Dawkins approves of "generosity and altruism."[9] Wilson wants to conserve the biosphere.[10]

Such atheistic moralism, however, raises the question: can these or any other values be sustained once religion is rejected? Can the Darwinism these thinkers profess justify the moral concerns they cherish? Wilson and Pinker, at least, venture to suggest that it can.

In *On Human Nature*, Wilson suggests that the progress of scientific knowledge will lead to the "dissolution of transcendental goals toward which societies can organize their energies." It will then become necessary and possible "to search for a new morality based upon a more truthful definition of man," to "look inward, to dissect the machinery of the mind and to retrace its evolutionary history." Human morality, he suggests, can be understood scientifically, as having been "programmed to a substantial degree by natural selection over thousands of generations."[11]

Similarly, Pinker challenges Pope John Paul II's view that the "dignity of the person" cannot be grounded on any theory of man that denies the spirituality of the soul.[12] On the contrary, he contends, there is a viable alternative "to the religious theory of the source of values": namely, that "evolution endowed us with a moral sense, and we have expanded its circle of application over the course of history through reason (grasping the logical interchangeability of our interests and others'), knowledge (learning the advantages of cooperation over the long term), and sympathy (having experiences that allow us to feel other people's pain)."[13]

This book tests the plausibility of such an evolutionary ethics by examining the work of those political scientists who have made sustained attempts to erect a normative or morally prescriptive political theory on the basis of Darwinian biology. In the last fifteen years a number of books by serious scholars have embraced a "new Darwinian naturalism in political theory" and asserted the possibility of a "Darwinian natural right" teaching—to use the expressions of one of this movement's most ardent and able proponents.[14] At the same time a number of political commentators have contended that the Darwinian account of human nature provides support for the prescriptions of political conservatism. This book challenges these positions. I contend that Darwinian biology cannot satisfactorily support morality, and accordingly that modern society cannot, contrary to the claims of evolutionary moralists and Darwinian political theorists, dispense with religion as a source of moral instruction.

Although this book is the fruit of my professional intellectual pursuits, it also has a personal interest for me, one that may clarify its relevance to those not necessarily concerned with the theoretical speculations of political scientists and conservative intellectuals. I must confess to holding some of those beliefs that Richard Dawkins thinks "no intelligent person" could reasonably entertain, and that I have held them on the understanding that they are *more reasonable* than the evident alternatives. In fact, I have been and continue to be influenced by C.S. Lewis's suggestion that "right and wrong" are a "clue to the meaning of the universe," that man's capacity for morality is a sign of his uniquely divine origins and, hence, of the insufficiency of any merely materialistic account of man.[15]

That belief was powerfully challenged for me when I took a graduate seminar in "Evolution and Political Theory" and was confronted with the impressive insights into our moral and political nature that can be gleaned from contemporary biology. Perhaps, it seemed, our understanding of right and wrong is not necessarily a clue to the meaning of the universe. Perhaps it can be understood sufficiently as a product of an evolutionary process driven by necessity and chance. My investigation of the new Darwinism in political theory was driven by my need to confirm for myself whether the evolutionary account had truly eliminated morality's apparent support for belief in man's uniqueness.

My conclusions are evident throughout the pages that follow: while I am still impressed with contemporary biology's insights into human nature, I do not think those insights can be made to add up to a sufficient account of our moral desires or, consequently, of who we are. Thus it seems to me that "right and wrong" may well still be taken as a "clue to the meaning of the universe." While my personal interest in this issue admittedly propelled my inquiry, I trust it did not compel my conclusions. Whether it did is for the open-minded reader to judge.

THE RIGHT DARWIN?

Darwinian Conservatism?

OES DARWINIAN BIOLOGY support conservative ideology? More specifically, can the Darwinian account of human nature lend scientific credibility to conservatism's moral and political prescriptions?

Until recently, such questions would have seemed absurd, at least to most conservatives. For until recently it seemed evident that conservatism should be hostile to Darwinian theory. Conservatives are thought to be, and think of themselves as, friends of public order, and in particular of the uncoerced public order that depends on the moral rectitude of the people. Yet a common conservative belief, going all the way back to Edmund Burke, holds that religion is one of the essential supports of such order, is in fact "the basis of civil society," because it alone can provide sufficient motives to ensure the willing good behavior of large numbers of people.[1]

Darwinism, however, insofar as it claims to explain the wonderful order of living nature as emerging, not from design, but from a combination of chance and necessity, appears hostile to religious belief and therefore, from a conservative standpoint, hostile to public morality. If, as Richard Dawkins has said, "Darwin made it possible to be an intellectually fulfilled atheist," then he struck a blow against the intellectual respectability of religion, a blow that would have to diminish its popular appeal.[2] This effect, no doubt, would be circuitous. Most people's faith is, if not derived from, at least supported by the example of those held in esteem by society. Thus Darwinism undermines the popular credibility of religion not directly, but by undermining its plausibility to intellectuals and scientists, who cannot then conceal their skepticism from the public.

This is not to say that Darwinism provides a refutation of religious belief comprehensible only to the scientifically literate and philosophically astute. Such a refutation is beyond the competence of science. It does, however, make belief in God appear much less a requirement of reason, and therefore much more a matter of pure faith, than it did before; and this must diminish the receptivity to faith of those who are inclined to give assent only in the presence of compelling evidence. This in turn fosters confusion and doubt among ordinary people, who cannot help noticing the apparent tension between two intellectual authorities for which they have a healthy respect: religion and science.

Conservatives have opposed Darwinism not only out of a desire to preserve the traditional morality they believe can be sustained only by a belief in God as the cause of the natural order, but also out of a fear of the novel morality they believe arises from what Darwinism posits as the real cause of the natural order: the process of natural selection. That is, they oppose Darwinian biology as a supposed cause of social Darwinism, which promotes a morality of unfettered and ruthless competition. As Darwinism contends that species develop through competition to survive and reproduce, some have concluded that society ought to model itself on nature—that by fostering, or at least doing nothing to moderate, ruthless competition we can foster the further development of the human race. The failure of the "less fit" to survive and reproduce, some have thought, means that in the future society will be more and more made up simply of the "fit": charity guarantees the continuation of those who rely on charity, while its absence, allowing nature to take its course, guarantees their gradual disappearance and the emergence of a society made up only of healthy and productive citizens. While such notions might be appealing to some elements on the right, they elicit the repugnance of traditional conservatism, which has always harbored a spirit of *noblesse oblige*, also traceable to Burke. One might say, then, that conservatives are inclined to reject Darwinism in the first place as a threat to the morality that should guide the people and in the second place as a threat to the morality that should guide elites.

Of course, not all conservatives, especially in America, adhere to the high-toned, philosophic conservatism of the intellectual descendents of Burke. Even if one embraces a more populist conservatism, however, Darwinism must appear no less threatening. After all, in America religious belief is popular: it is of the people, a widespread phenomenon. Therefore, ordinary citizens view Darwinism's apparent hostility to religion with alarm, not so much because they think that religion is a necessary support for their own decency, but more because faith in God is simply a cherished part of their psychological equipment. At the same time, populism is supposed to be on the side of the people, as opposed to the elite; and the people in this sense are those who are economically insecure: if not poor themselves, they are at least close enough to poverty to see it and to feel vulnerable. Thus a populist conservatism can have no sympathy for social Darwinism, which encourages elites to act as if they have no obligation to the less fortunate.

DARWINIAN CONSERVATIVES?

Today, however, the question of the proper conservative stance toward Darwinism is more problematic, for in recent years a growing number of conservative intellectuals have advanced the claim not only that

Darwinism and conservatism need not be antagonistic, but even that they can and should be allied. The compatibility of Darwinian theory with conservative moral and political principles has been advanced in popular organs of conservative opinion like *First Things* and *National Review*. This new Darwinian conservatism has been made possible by the rise of sociobiology or evolutionary psychology—the scientific exploration of the biological basis and evolutionary origins of the behavioral and mental propensities of human beings—and its emphasis on, and apparent restoration of the credibility of, human nature. The focus of earlier Darwinian theorizing, or at least that which raised the concerns of conservatives, was not human nature, but nature viewed on a grander scale, either cosmic or biological. This part of Darwinian theory proceeded according to assumptions that seemed indirectly to undermine morality. The Darwinian cosmos was thought to be ruled by chance and necessity, and thus to leave no room for a designer-God who might legislate morality for men. And Darwinian living nature was thought to be governed solely by competition to survive and reproduce, and thus to provide no support for moral obligation as distinct from self-interest.

Now, however, Darwinian scientists have turned their attention to human nature, and the results are said to be comforting to conservatism. Contrary to the denials of some on the left, human nature, according to evolutionary biology, exists and is relevant as a principle limiting and guiding moral and political choice—limiting, in fact, the aspirations of the left and guiding us in the direction of conservatism. Hence, for example, Charles Murray's claim that "the story of human nature as revealed by genetics and neuroscience" is so far, and will likely continue to be, "Aristotelian in its philosophical shape and conservative in its political one."[3] More specifically, Larry Arnhart urges conservatives to "see that Darwinian views of human nature provide scientific support for the traditional idea of natural moral law" and for the conservative notion that human beings are "naturally social and moral animals," rather than, as the left would have it, autonomous creators of values.[4]

In this light, the older conservative concerns about social Darwinism as an ideology of ruthless and amoral competition disappear. It may be true that evolution is driven by such an amoral process, but it happens to have generated a human nature that is moral and that includes impulses to sympathy and fair-dealing. And it is our nature as we experience and know it, and not the process by which it emerged, that is normative for us. Thus, regardless of the character of natural selection, social Darwinism is unnatural, in the sense of being contrary to *our* nature. Evolution may be a mindless and amoral process, but it does not follow that human beings should or do value mindless amorality.

The sociobiological recovery of human nature also tends to alleviate the concern that Darwinism is hostile to morality because it undermines the credibility of religion. Even if Darwinian science does call religion into question, the moral problems arising from this are self-correcting: insofar as Darwinism reveals a scientific basis for morality, religion becomes superfluous. Why, as Arnhart asks, rely on myth as a support for morality when science is available and suitable to the task?[5]

In addition to highlighting the general harmony between conservatism and Darwinism, such thinkers claim evolutionary biology's support for conservative positions on a range of particular issues. Murray contends that as our scientific knowledge of human nature becomes ever more precise "the adages of the right will usually prove to be closer to the mark than the adages of the left, and many of the causes of the left will be revealed as incompatible with the way human beings are wired."[6] More precisely, modern evolutionary science asserts the natural status, and hence both the ineradicable character of and the justice of, the traditional family,[7] of sex-based psychological differences that point to different and traditional sex roles,[8] of the possession of property,[9] of the reciprocal exchange that occurs in markets,[10] and of intellectual inequalities that generate inequalities in achievement.[11]

Indeed, the Darwinian conservatives contend that evolutionary science ultimately promises (or threatens) not merely to refute the liberal agenda one issue at a time, but fundamentally to delegitimize the entire outlook of the left, which emphasizes the transformation of man through the reinvention of the social environment. Such an aspiration depends on the assumption that the social environment is the decisive force shaping human life—an assumption that modern evolutionary biology denies by its recovery of human nature, of innate or biologically based behaviors and patterns of thought and feeling.[12] This explains, conservatives charge, the hostility of the left to much modern evolutionary science.[13] Indeed, the academic left has from the beginning sought to discredit even the most theoretical and apolitical presentations of this new science, such as E.O. Wilson's *Sociobiology*.[14]

In addition to such public arguments for the compatibility of Darwinism and conservatism there is a growing body of scholarship in political science that Larry Arnhart has called the "new Darwinian naturalism in political theory."[15] This literature, too, lends support to the conservative notion that morality generally is rooted in human nature and not just a product of social construction, as well as to more specifically conservative beliefs in, for example, the natural status of the family and of traditional sex differences. Indeed, much of this scholarship has been authored by academics who may fairly be regarded as conservatives, such as Arnhart himself, James Q. Wilson, and Francis Fukuyama.

Despite the growing numbers of, and undeniable intellectual seriousness of, those offering Darwinism as a scientific basis for conservatism, many conservatives are still not buying. While sociobiology is controversial, in the sense of being hated, on the left, it is controversial, in the truer sense of being disputed, on the right. Thus the proposed alliance between Darwinism and conservatism seems to have been blamed as often as it has been praised in the pages of conservative periodicals. Many conservatives, it seems, find the Darwinian account of human nature at best useless and at worst pernicious.

Some contend that, because the Darwinian account of human nature establishes the natural status of such a diverse set of often conflicting human propensities, it can be used to support a wide variety of moral and political positions, and therefore is worthless as a support of any particular ideology.[16] Other critics object to what they take to be the implicit consequences of Darwinism's presentation of morality as a product of the natural striving for survival and reproduction. Darwinism, they suggest, delegitimizes (because it cannot explain) generous moral impulses that transcend kinship and reciprocity, and denies as superfluous the notion of God as the basis of morality.[17] These critics also object to Darwinism's materialism or "naturalism"—its contention that the operation of merely material forces can explain everything in nature, even the constitution of the human mind—and suggest that such an account necessarily erodes moral responsibility because of its implicit denial of the reality of free will.[18]

TOCQUEVILLE'S CONSERVATISM

How, then, should conservatives view this attempt to harmonize Darwinism and conservatism? This question cannot be answered without some effort to posit what conservatism should seek to conserve. I propose to answer this question, and to evaluate the new Darwinian political theory, from the standpoint of the conservatism of Alexis de Tocqueville, a conservatism that seeks to conserve the conditions of human dignity and freedom within modern democracy.

Tocqueville was no aristocratic reactionary, and he certainly appreciated the benefits to be derived from the democratic social state that he foresaw would rule the modern world. Democracy, he conceded, is more just than aristocracy, insofar as the former, unlike the latter, does not confine men to the social and economic position into which they happen to be born. Moreover, democracy, through the ambitious energy liberated by its promise of social mobility, generates a much greater and more widely distributed prosperity than aristocracy. Nonetheless, Tocqueville also recognized in democracy certain dangerous tendencies in need of restraint, tendencies opposed to human flourishing and to justice. Hence his attempt, in *Democracy in America*, to elaborate a "new political science" for a new, democratic world, a science of statesmanship able to "regulate" democracy's "movements" and to replace its "blind instincts" with "knowledge of its true interests."[19]

In particular, Tocqueville feared that democracy fosters a popular obsession with physical comforts unworthy of our humanity and that it dangerously inclines to majority tyranny. He emphasized religion's role in restraining such democratic tendencies, and therefore *Democracy in America* accords religion a prominent place in Tocqueville's new science of politics and in the cultural equipment of a healthy democracy. Thus, he insists on the compatibility of religion and democracy, and even suggests that "one must maintain Christianity within the new democracies at all cost."[20]

In this book, then, I seek to evaluate the Darwinian account of human nature and morality in light of Tocqueville's conservatism. I ask, can Darwinism provide the moral instruction and restraint that democracy needs? Can modern democracy, contrary to Tocqueville's expectation, sustain human dignity and freedom without the support of religion? I explore these questions by examining Tocqueville's account of democracy's moral weaknesses, his understanding of the possibility of a religious correction of them, and the arguments of the proponents of the new Darwinian political theory, in particular, Larry Arnhart, Francis Fukuyama, James Q. Wilson, Roger Masters, and Robert McShea.

SCIENCE, MORALITY, AND FAITH

One might well object to such an enterprise on the grounds that it subjects science to an inappropriate moral or political test. Science aims to uncover the truth, and it is not fitting or pertinent to ask whether the truths discovered are helpful or deleterious to some political or moral aspiration. Such an inquiry as this book undertakes, however, is legitimate on its own terms, even if those terms are not primarily scientific. The discovery of scientific truth is certainly a legitimate and even admirable enterprise, but the psychological and moral needs of society are no less worthy of our concern. That something is scientifically true does not necessarily mean that it will be politically or morally useful to know, or even that it will not be somehow harmful in some social conditions. And because the political and moral health of a society is as important as the truths that science reveals, the extent to which those truths are compatible with such health is as worthy of investigation as those truths themselves. In other words, that a scientific account of things is morally or politically problematic for a given society is not itself a refutation of that account, but neither is it for that reason irrelevant.

Moreover, the proponents of the new Darwinian political science themselves invite such an inquiry by making practical or moral claims on behalf of that science. That is, they present it not only as theoretically true but also as publicly salutary, as a more wholesome alternative to influential modern ideologies corrosive of moral seriousness. Thus, for example, James Q. Wilson's *The Moral Sense* aspires not only to demonstrate the natural status of our moral inclinations, but also to "help people recover the confidence with which they once spoke about virtue and morality."[21] Wilson begins his study by deploring the effects of popular moral skepticism, which tends to excuse the actions of the least moral among us and which makes impossible "any serious discussion of marriage, schools, or mass entertainment." He hopes, in response, to rebuild the basis of our moral judgments.[22] Similarly, Francis Fukuyama's *The Great Disruption* both notes the possible deleterious consequences of cultural relativism for a democratic society and points out hopefully that modern evolutionary science calls cultural relativism into question.[23] And Roger Masters, in *The Nature of Politics*, offers an evolutionary account of human nature that can, "unlike the nihilism and relativism that have predominated in the West over the last century," provide the basis for "decent and humane standards of social life."[24] Thus a political or moral critique of the new Darwinian political theory is not out of place because that theory presents itself not only as science but also as a public philosophy—or at least as a basis for one.

Moreover, the inquiry this book undertakes is invited not only by the Darwinian political theorists' general assertion of their science's moral benefits, but also by their more specific contention that their evolutionary account of morality renders religion superfluous as a support for moral order. Tocqueville argues that "dogmatic beliefs in the matter of religion" are "most desirable" for society, because "[t]here is almost no human action . . . that does not arise from a very general ideal that men have conceived of God, of his relations with the human race, of the nature of their souls, and of their duties toward those like them."[25] In contrast, the Darwinian political theorists hold that morality can flourish even in the absence of religion, because moral behavior, being rooted in our natural passions, does not fundamentally depend on beliefs about metaphysical questions such as the nature or existence of God or the soul.

Wilson, for instance, contends that social order requires no "fear of eternal damnation" but instead can rest sufficiently on "the instincts and habits of a lifetime, founded in nature, developed in the family, and reinforced by quite secular fears of earthly punishment and so-cial ostracism."[26] Similarly, Robert McShea, in *Morality and Human Nature*, devotes part of a chapter to debunking belief in God as a basis of morality and concludes that "[b]elievers and non-believers seem to have about equal motivation to behave well toward fellow humans."[27] And Larry Arnhart, in *Darwinian Natural Right*, holds that, while religion can admittedly "*reinforce* the moral sense," "insofar as the moral sense is natural, it does not *require* religious belief."[28] Thus, while Tocqueville's political science finds that the necessity of religion "may be deduced very clearly even if one wants to pay attention only to the interests of this world,"[29] the new Darwinian science of politics assures us "that our earthly happiness is securely founded in our nature as moral animals endowed with a moral sense that serves our natural desires."[30]

WHY TOCQUEVILLE? WHY CONSERVATISM?

Even if one grants the propriety of evaluating the moral and political implications of the new Darwinian political science, however, questions remain regarding the manner of the investigation attempted here. In the first place, conservatives might well ask why they should take their bearings from Tocqueville in judging what Darwinism has to offer. One might point out in response that Tocqueville is a useful guide because he avoids two opposite, immoderate temptations of traditional conservatism: nostalgia and cynicism. Conservatism looks to the past for guidance, and while there is much to recommend such a disposition, it also involves the peril of waxing nostalgic for a supposedly pure bygone era. Thus conservatives sometimes fall prey to the very vice for which they criticize the left, utopianism, with the difference that they find their utopia in the past rather than the future. Tocqueville's analysis, however, combines a healthy respect for the achievements of the past with a healthy realism about what can be achieved in the present and future. He recognizes the virtues of aristocracy, especially its elevated conception of human possibilities, without romanticizing it, and without pretending that those virtues can be replicated in democracy. Rather, he limits himself to using the excellences of the feudal past to illuminate democracy's deficiencies and hence to provide a basis for aspirations that are realistic while also going beyond the morality that untutored democracy is likely to provide.

Tocqueville's realism, however, does not go so far as to become cynicism, which is the other, opposite temptation to which conservatives are subject. Seeking to avoid the errors of the left, some conservatives pride themselves on their realism, their immunity to utopian dreams. Such a disposition is in danger of simply surrendering to powerful moral and cultural trends, regardless of their wholesomeness. After all, to resist them, thereby courting political marginalization, is the mark of the fanatical ideologue rather than the prudent conservative. Tocqueville, on the other hand, while realistically conceding that democracy possesses powerful and dangerous inclinations that cannot be eradicated, still insists that public-spirited citizens and statesmen are obliged to try to restrain and educate them.

Populist conservatives are certainly not tempted to cynicism or to an undue idealization of the past. They are perhaps tempted, however, to idealize the moral and political opinions of ordinary citizens and therefore the democratic society in which those opinions tend to govern public life. This explains their tendency to view Darwinism, on the one hand, as simply a corrupting principle emanating from the irresponsible speculations of impious intellectual elites, while viewing the common sense opinions of the people, on the other hand, as unproblematic as the basis of a decent public order. From Tocqueville such populists might learn, as we will see, that democracy shapes common opinion in some ways that are deeply problematic.

Moreover, populists might be tempted to assume the easy compatibility of the things they love, the things Americans cherish, especially
equality and religion. Tocqueville, however, suggests that while societies founded on equality can be improved by religion, they are not
necessarily friendly to it. And in recognizing the problematic status
of religion in democracy, populist conservatives might see how it is
at least plausible that contemporary Darwinism could present itself
as a morally salutary public philosophy.

What do I have to offer the non-conservative? Certainly Tocqueville's concerns with democratic materialism and tyranny of the
majority are sufficiently broad as to be of interest as well to liberals
and those who subscribe to no particular ideology. Whether religion
is necessary to democracy's flourishing, or whether democratic men
can now receive sufficient moral instruction from science, is a question of similarly wide interest. Moreover, given the public authority
of modern science, whether and to what extent it really sustains a
conservative understanding of society should be of interest no less
to the liberals whose rhetorical position is weakened by such claims
than to the conservatives who hope to demonstrate the scientific
basis of their prescriptions.

In any case, conservatism and Tocqueville's account of democracy
are the points of departure, but not the exhaustive framework, for
the argument that I offer in the following chapters. I hope to show
that the new Darwinian political theory is insufficient not only from
the standpoint of conservative and Tocquevillian concerns, but also
from the standpoint of the moral concerns of any human being of
whatever ideological persuasion. Ultimately, I contend that the new
Darwinian political theory is inadequate even according to the moral
aspirations of its own proponents—and indeed that it provides the
basis for no useful moral teaching at all.

Before proceeding with this critique, I should offer the following
caveat, made necessary by the highly charged atmosphere in which

the merits of evolutionary theory is typically debated. To question Darwinism's ability to provide a sufficient justification for morality, as I do in the following chapters, is not necessarily to deny the reality of evolution according to natural selection. On the other hand, to concede that the evolutionary account offers useful insights into our nature, as I also do in the following chapters, is not necessarily to concede that it can completely account for the fullness of our humanity. These points should go without saying, but in an intellectual environment in which charges of both fundamentalism and atheism are thrown about too freely, they perhaps need to be said.

Decent Materialism

WHILE TOCQUEVILLE FREELY ACKNOWLEDGES the blessings that flow from modern democracy—the freedom of social mobility, the energy it unleashes, and the prosperity it generates—he also sees in it certain tyrannical tendencies that pose a grave threat to human dignity and freedom. On the one hand, democracy encourages hedonism and thus fosters a tyranny within each citizen, a despotism of the worse, less human elements of the soul over the better and more human. On the other hand, democracy tends toward the establishment of tyranny in the society at large, specifically the tyranny of the majority, which subordinates the rights of minorities to the interests of the dominant faction. Tocqueville finds in religion, with its emphasis on the good of the soul and its insistence on moral obligation, a possible corrective to these dangerous democratic tendencies. Nevertheless, his account also suggests doubts about the power of religion's hold on the democratic mind. It thus points to the possibility that democracy's moral instruction will have to be provided by science rather than faith, and accordingly seems to invite the efforts of those who would seek moral knowledge from Darwinian evolution rather than from Biblical revelation.

DEMOCRATIC MATERIALISM

"In America," Tocqueville observes, "the passion for material well-being," while not "always exclusive," is at least "general." "The care of satisfying the least needs of the body and of providing the smallest comforts of life preoccupies minds universally." This phenomenon, however, arises not from any peculiar vulgarity of the American character but from the nature of democracy itself. For, while the "taste for material well-being" is "natural and instinctive," and therefore to be found in all societies, democratic social conditions excite this passion to an extent unheard of in aristocratic nations.[1]

An aristocratic social state, Tocqueville contends, limits the intensity of this natural longing by offering to some an easy satisfaction of it and by imposing on others its near-complete denial. In an aristocracy, the rich, born to a life of secure privilege and comfort, have no knowledge of, have no fear of, and can "hardly imagine" any other state of life. Physical comforts therefore become for them "not the goal of life" but merely "a manner of living." They enjoy such pleasures, but without thinking much about them; and "their souls transport themselves elsewhere," pursuing "more difficult and greater" enterprises, perhaps the challenges of ruling or the satisfactions of artistic, literary, or philosophic achievement.

Shifting his attention to aristocracy's lower classes, Tocqueville finds "analogous effects produced by different causes." While aristocrats are freed from hedonism by the familiarity and security of physical comforts, for the people such pleasures hold little interest because of their unavailability. Where social mobility is impossible, "the people" become as accustomed to "poverty" as "the rich" become to their "opulence." Thus the former give little thought to material well-being "because they despair of acquiring it and because they are not familiar enough with it to desire it." As a result, the people achieve a certain elevation of mind. Confronted with such unpromising material prospects, their "imagination . . . is thrown back upon the other world," where it seeks enjoyments beyond the "miseries" of this life.[2]

Tocqueville's account rests on the psychological principle that the "human heart" is attached most "keenly" to a "precious object" not by its "peaceful possession," nor by its simple unavailability, but by its partial and insecure acquisition, which is accompanied by "the imperfectly satisfied desire" to have it and the "incessant fear of losing it." Democracy, then, so powerfully attaches the heart to material comforts because, while its openness to social mobility places such comforts within the reach of anyone, the competition that it unleashes renders the possession of such comforts uncertain for all. Democracy, by destroying privileges and dividing inheritances, establishes a "multitude of mediocre fortunes." Democratic men therefore "have enough of material enjoyments to conceive the taste" for them but "not enough to be content with them. They never get them except with effort, and they indulge in them only while trembling."[3]

Tocqueville identifies a number of disturbing consequences that follow upon this democratic materialism. The susceptibility of some Americans to religious fanaticism, for example, results from the worldly preoccupations of the dominant culture. Because the love of the "infinite" and "immortal" is rooted in human nature, Tocqueville argues, the neglect of such needs predictably renders the soul "bored, restive, and agitated among the enjoyments of the senses." Thus when the majority is concerned only with "the search for material goods," it is not surprising that "an enormous reaction" is "produced in the souls of some men," who "throw themselves head over heels into the world of spirits" and in their piety often go "beyond the bounds of common sense."[4]

On the other hand, most Americans, those who steadily avoid such spiritual enthusiasms and devote themselves primarily to the goods of the body, seem to Tocqueville "grave and almost sad even in their pleasures." A pronounced "restiveness" accompanies this sadness, and arise, again, from democratic materialism. The man who restricts himself to the pursuit of worldly comforts "is always in a hurry," constantly prodded by his imagination of the many goods "that death will prevent him from enjoying if he does not hasten." Thus his soul is filled "with troubles, fears, and regrets" and a "sort of unceasing trepidation that brings him to change his designs and his place at every moment."[5]

In addition, Tocqueville warns that democratic materialism opens the door to a kind of despotism by distracting citizens from politics. To those whose "taste for material enjoyments" has progressed more quickly than their "enlightenment" and habituation in freedom, and who consequently are concerned only with the acquisition of wealth, the execution of civic obligations seems merely a "distressing contretemps that distracts them from their industry." "There is," Tocqueville writes, "no need to tear from such citizens the rights they possess," for they "willingly allow them to escape." If, under such circumstances, "an ambitious, able man comes to take possession of power, he finds the way open to every usurpation." Alternatively, in such conditions society may fall prey to the "despotism of factions." "When the mass of citizens wants to be occupied only with private affairs," Tocqueville warns, "the smallest parties should not despair of becoming masters of public affairs," and one may then be "astonished at seeing the small number of weak and unworthy hands into which a great people can fall."[6]

Most fundamentally, however, Tocqueville regards democratic materialism as harmful not only because of these consequences, also but because it is itself degrading and unworthy of human nature.

Tocqueville is careful to specify the character of this degradation: it is not depraved, but base. There is little reason to fear, he contends, that democratic materialism will manifest itself in "debauchery," in a "sumptuous depravity and a brilliant corruption." On the contrary, Tocqueville indicates that such vices belong not to ordinary democrats but to corrupt aristocrats—those who, having lost their authority but retained their wealth, with the capacity to do much but with nothing to do, apply their considerable imagination, ingenuity, and resources to the satisfaction of the body. Democratic men, on the other hand, are careful to avoid physical pleasures that are contrary to "order" or to "regular mores." Presumably both the small scale and the insecurity of the fortunes typically made in democracies render the possessors cautious in their indulgence of pleasures, lest in the pursuit of some self-forgetting ecstasy they lose everything. The democratic man's imagination, then, dwells not on the possibility of "depleting the universe" to satisfy his own desires, but instead on humbler matters such as "enlarging" his home, "making life easier and more comfortable at each instant, preventing inconvenience, and satisfying the least needs without effort and almost without cost."[7]

For Tocqueville, such democratic hedonism, despite its careful avoidance of outright immorality, is nonetheless degrading because, by dominating the mind, it drives out other pursuits more worthy of our humanity. While the objects the democratic soul seeks are in themselves permissible, "it clings to them" with an unseemly tenacity. The democratic mind contemplates such petty pleasures "every day and from very close," and Tocqueville fears that "in the end" they will shut out the "rest of the world" and even come between the soul and God. Thus in "striving to seize" legitimate material benefits, democratic men tend to forget "the more precious goods that make the glory and greatness of the human species."

TYRANNY OF THE MAJORITY

Tocqueville approves of democratic government, the rule of the majority, but he fears its abuse, the tyranny of the majority. While the majority of a people should ordinarily make law for a particular society, he contends, it ought not do so in violation of "justice," the "law" of "universal society." Thus the "sovereignty" of any particular people must be subordinate to the "sovereignty of the human race." For Tocqueville, moreover, abusive deviations from universal justice are a real possibility in democracy. A majority, he notes, may reasonably be viewed as "an individual" possessing "opinions" and "interests" conflicting with those of "another individual," namely the "minority." Tocqueville sees no reason to think that collections of men are less likely than one man to abuse their power over their adversaries.[8]

In fact, one can argue that, on Tocqueville's account, majorities are inclined toward tyranny not only by the imperfect, self-interested human nature of their individual members, but also by the convergence of two characteristics of the equality that is both the foundation of the democratic social state and the democrat's most revered principle. First, the equality established by democracy constricts the perceived scope of moral obligation within a very confined sphere, or, as Tocqueville puts it, fosters a spirit of "individualism" among the citizens. While aristocracy's hierarchy of rank necessarily creates a system of mutual dependence and obligation binding all individuals, democracy "breaks" this "chain and sets each link apart." Under democracy's conditions of equality and the social mobility that they permit, "one finds a great number of individuals who" are not wealthy enough to command much influence over others but sufficiently wealthy to be independent. Such men "owe" and "expect" nothing "from anyone." Thus democracy "disposes each citizen to isolate himself from the mass of those like him," to "withdraw to one side with his family and friends," and "threatens finally to confine him wholly in the solitude of his own heart."[9] Tocqueville draws no explicit connection between this democratic individualism and the possibility of majority tyranny, and in the context his emphasis is on the troubling extent to which such individualism actually renders citizens indifferent to politics. Nevertheless, it seems reasonable to fear that, under the individualistic conditions Tocqueville describes, whenever some common interest unites the members of any majority it will be unchecked by any deeply felt sense of obligation to the members of the opposing minority.

Second is the democratic belief in the superiority of the opinions, interests, and rights of the majority, which further aggravates the vulnerability of the minority within democracy. Tocqueville notes that democratic peoples tend to submit not only to the majority's legal power but also to its "moral empire." They concede not only the majority's right to direct society but also come to accept as correct whatever it wills. This moral empire, Tocqueville contends, arises from the democratic experience of and belief in equality. Aristocracy fosters submissiveness to intellectual authority by providing stark examples of intellectual superiority, which stem from the restriction of leisure and learning to a small segment of society. In contrast, democracy, by destroying inherited privilege and thus rendering men equal in their access to education, produces a rough equality of intellect that in turn renders each man little disposed to trust the judgment of any other individual. This same equality, however, increases the proclivity to trust the judgment of the majority. "In times of equality, because of their similarity, men have no faith in one another; but this same similarity gives them an almost unlimited trust in the judgment of this public; for it does not seem plausible to them that when all have the same enlightenment, truth is not to be found on the side of the greatest number."[10]

Democracy, then, fosters submissiveness to public opinion on the part of those outside the majority and, on the part of those who are themselves members of the majority, an unshakable (though irrational) confidence in the justice of their position. Such proclivities must conspire to render minorities vulnerable to oppression. Thus Tocqueville notes the denigration of individual rights that unchecked democracy tends to generate: "As conditions are equalized in a people, individuals appear smaller and society seems greater," for "each citizen, having become like all the others, is lost in the crowd, and one no longer perceives [anything] but the vast and magnificent image of the people itself." Consequently, men living in democratic times "naturally" entertain a "very high opinion of the privileges of society and a very humble idea of the rights of the individual," taking the interests of the former to be "everything" and those of the latter to be "nothing."[11]

THE BENEFITS AND LIMITS OF RELIGION

Tocqueville presents religion as a necessary corrective of these dangerous proclivities and hence a necessary support for human dignity and freedom in democratic times. Having described the American manifestation of democratic materialism, he points to worship as the only activity capable of distracting Americans from their almost single-minded pursuit of comfort. On Sunday, he observes, commerce and industry cease, and Americans go to church, where they are "told of the necessity of regulating" their "desires, of the delicate enjoyments attached to virtue alone, and of the true happiness that accompanies it." Returning home, they open the Bible to consider "sublime or moving depictions of the greatness and goodness of the Creator, of the infinite magnificence of the works of God, of the lofty destiny reserved for men, of their duties, and of their rights to immortality." Thus do Americans escape the "small passions" that "agitate" their lives and discover "an ideal world in which all is great, pure, eternal."[12]

This influence of religion is salutary, on Tocqueville's account, not because it deters democrats from a chaotic criminality, to which their materialism is not inclined in any case, but because it saves them from a kind of orderly dehumanization, which is, again, the real danger posed by democratic materialism. The "whole art of the legislator," Tocqueville contends, "consists in discerning," under particular social circumstances, which inclinations require encouragement and which require restraint. In an aristocracy, which tends naturally to limit the love of physical comforts, one must "try to excite" the "search for well-being." Democratic peoples, however, will "perfect" without prompting "each of the useful arts and render life more comfortable, easier, milder every day." The "peril" of democracy, then, is that, while enjoying this "honest and legitimate search for well-being," men will in the end cease to make use of their "most sublime faculties" and ultimately "degrade" themselves. Therefore, Tocqueville holds, it is the obligation of democratic legislators, and indeed of "all honest and enlightened men," to "apply themselves relentlessly to raising up souls and keeping them turned toward Heaven," to foster "a taste for the infinite, a sentiment of greatness, and a love of immaterial pleasures."[13]

Religion also helps restrain the tyranny of the majority by recalling to democratic men—preoccupied with themselves, because of their independence, and dismissive of the rights of minorities and individuals, because of their equality—the universal principles of justice that unite all human beings and that bind even majorities in their conduct of society's affairs. While democracy tends to "isolate" men "from one another and to bring each of them to be occupied with himself alone," religion inspires "wholly contrary instincts" by imposing on "each some duties toward the human species or in common with it," thus drawing man, "from time to time, away from contemplation of himself." While equality suggests that the interests of society are everything and those of the individual nothing, thanks to Christianity, Tocqueville reports, no American (at least up to his time) has been found to assert "that everything is permitted in the interest of society," an "impious maxim . . . that seems to have been invented in a century of freedom in order to legitimate all the tyrants to come."[14]

In light of the gravity of the democratic defects that religion tends to correct, Tocqueville insists, contrary to the views of his liberal European contemporaries, that religion is not only compatible with democracy but necessary to its flourishing.[15] He includes religion among the "Principal Causes Tending to Maintain a Democratic Republic in the United States" and ultimately names it "the first" of America's "political institutions," pointing to its "singular" role in facilitating, by disciplining, the people's use of their power.[16] And he holds that the American belief that democracy must be made "more moral"—that is, less hedonistic—"by means of religion" is "a truth with which every democratic nation ought to be instilled." He goes so far as to contend that "one must maintain Christianity within the new democracies at all cost."[17]

Given these concerns, one might reasonably conclude that Tocqueville would look with disfavor on any theory that tends to undermine religion's intellectual credibility, and therefore that he would, given an opportunity, take a place among the ranks of the traditional conservative opponents of Darwinism. Indeed, this impression is strengthened by the fact that, in both his discussion of religion's role in moderating the tyranny of the majority and his later account of its ability to temper democratic materialism, Tocqueville makes a point of issuing rather forthright condemnations of implicitly atheistic philosophical or scientific systems. He notes, for example, the arguments of those who hold that "freedom and happiness" would be advanced by belief "in the eternity of the world" and in the notion that "the brain secretes thought." Tocqueville, however, dismisses such thinkers with the observation that they have no experience of either religious peoples or free ones. Elsewhere he contends that, because the belief "that everything perishes with the body" tends to give added force to the already excessive democratic preoccupation with physical well-being, philosophic materialists must be regarded as "natural enemies" of democracy.[18]

This account of Tocqueville's belief in the necessity of religion and his consequent hostility to scientific materialism, however, does not exhaust his understanding of religion's place in democracy. The rest of the story, paradoxically, points to the possibility of a Tocquevillian conservatism favorable to Darwinism. Tocqueville's account, it seems, emphasizes not only democracy's need for, but also its hostility to, religious belief. Faith, while important as a restraint on democracy's dangerous inclinations, does not readily grow in the soil of equality. Indeed, Tocqueville's analysis of democracy's influence on the minds of citizens suggests that it tends gradually to undermine religious belief. This difficulty leads one to wonder whether Tocqueville might embrace a Darwinian account of man that can plausibly be presented not as a threat to, but as an alternative basis for, the morality that Tocqueville thought could be provided only by religion.

Americans, Tocqueville observes, are natural Cartesians: in "most of the operations of the human mind, each . . . calls only on the individual effort of his own reason," and all tend "to seek the reason for things by themselves and in themselves alone."[19] This disposition is the fruit not of philosophic instruction but of an intellectual instinct fostered by the equality that characterizes the democratic social state. The key to aristocracy, Tocqueville contends, is the law of primogeniture, which "concentrates" first "property" and then "power" around "some head," spontaneously generating a society divided between those who must labor and those who need not. In contrast, when the law of inheritance commands or even permits "equal partition of the father's goods among all the children," estates are "constantly fragmented into smaller portions," creating a society devoid of both compulsory poverty and hereditary leisure. Under such democratic conditions, social advancement is open to all, but all must work to make progress or to maintain such prosperity as they have achieved. All have access to some measure of (mainly practical) education, but there is no permanent class with leisure for a refined education undertaken for its own sake.[20] The effect of democracy, then, in contrast to aristocracy, is to compress the range of differential intellectual achievement. Thus Tocqueville finds America the country with fewer "ignorant" and fewer "learned" men than in any other," where "a common level in human knowledge has been established," to which "[a]ll minds have approached," some by being "raised" and others by being "lowered."[21]

While aristocracy, by creating vast differences in education, disposes people to respect intellectual authority, democracy, by leveling those differences, destroys that disposition. Respect for intellectual authority, Tocqueville notes, "is necessarily very restricted in a country where citizens" are "nearly the same" and therefore see in no one "incontestable signs of greatness and superiority." In the absence of obvious intellectual inequality, citizens are "constantly led back toward their own reason as the most visible and closest source of truth."[22]

Such democratic rationalism, however, cannot but gradually erode religious belief. Habituated to relying only on their own understanding, and doing so, at least in their everyday affairs, with some success, democrats begin to believe that reason is equal to explaining everything and therefore begin to doubt whatever surpasses their understanding. As a result they tend to "conceive a sort of instinctive incredulity about the supernatural," and while they recognize that some intellectual authority is necessary in any society—hence, again, their deference to the views of the majority—they are "only with difficulty" led to place such authority "outside of and above humanity."[23] Moreover, because of the absence of leisure characteristic of the democratic social state, most people will be "more occupied with business than with studies, with political and commercial interests than with philosophical speculations or belles-lettres." Therefore democracy's language itself will "abandon little by little the terrain of metaphysics and theology."[24] Finally, then, on Tocqueville's account religion appears not as part of democracy's natural cultural equipment but instead as a kind of "inheritance from aristocratic centuries."[25] And perhaps it is in light of this analysis that he observes that even in America, which he praises for its popular religiosity, religion commands respect "much less as revealed doctrine than as common opinion."[26]

THE ATTRACTION OF DARWINISM

Initially presented as a solution to democracy's problems, Tocqueville's new science of politics is ultimately revealed as in some measure problematic itself. It offers religion as a restraint on some of democracy's most dangerous inclinations, yet it also shows that democratic inclinations tend to weaken religion. In the face of democracy's simultaneous need for and resistance to religion, we are led to wonder whether something else can fulfill the moralizing functions performed by religion in Tocqueville's political science. Here the new Darwinian political theory, with its claim to provide a scientific basis for human morality, presents itself as a plausible alternative.

Darwinism, after all, can harmonize with the democratic imagination in ways that religion cannot. Most obviously, as a product of modern science, the Darwinian account of morality appeals to rather than repels the democratic rationalism that Tocqueville observes. In addition, Darwinism responds to the democratic taste for what Tocqueville terms "general ideas." While the very structure of aristocracy impresses upon the mind the importance of distinctions and differences, Tocqueville notes, the absence of privilege in democracy leads to, and creates expectations of, a certain homogeneity among human beings. Thus, when considering society, democrats tend to think in terms not of ranks and classes but of human nature itself. The democratic man, however, "carries this same habit" of thought into all other realms of inquiry, and the instinct for general ideas, the desire to "explain a collection of facts by a single cause," becomes "an ardent and often blind passion of the human mind."[27]

The Darwinian account of the origins and functions of morality, however, responds to this democratic desire. For Darwinism holds that not only the bodily articulation of species, but also all animal instincts, including our moral and sociable emotions, can ultimately be traced to the operation of natural selection. Not only, say, the opposable thumb, but also our capacity for sympathy or feelings of concern for kin, were favored by evolution because of their contribution to the survival and reproduction of our ancestors. Ultimately, on the Darwinian account, all of human nature, and all of living nature itself, from the crudest physical capacities to the most refined workings of the human intellect, can be traced to the operation of natural selection. Thus does Darwinism appeal to the democratic longing to "link a multitude of consequences to a single cause."[28]

Here some clarification is in order, as this remark on the strength of Darwinism as a public philosophy could be interpreted as an implicit criticism of its quality as science. The point of these observations is not to suggest that the Darwinian account of morality is merely an expression of minds enslaved to the simplifying habits of thought fostered by equality of conditions, nor is it to deny the real complexity and subtlety of Darwinian explanations of human capacities and behavior. Whatever the role of democratic psychology in predisposing modern scientists to such theories, the theories themselves must be evaluated on their own merits. And because Darwinian explanations recognize the role of particular and variable environmental circumstances in influencing which capacities will be ratified as adaptive by natural selection, such explanations cannot, as their proponents point out, justly be presented as crudely deterministic or reductionistic. Rather, my only point here is that by its *surface* simplicity, and perhaps even its *apparent* reductionism, the Darwinian account of morality lends itself to a popularization that appeals to the democratic taste for general ideas. Hence weekly news magazines' frequent summarizations of evolutionary psychology and sociobiology.

Darwinism, then, seems to possess for the democratic mind the prima facie plausibility that religion lacks. If Darwinism is strong where religion is weak, however, it must also, if it is to supply the functions of religion in Tocqueville's political science, be strong where religion is strong: it must be able to justify a morality that can restrain democracy's destructive inclinations. At least at first sight, the new Darwinian political theory seems capable of this. To the extent that this science reveals our moral impulses as no less natural than our bodily desires, it appears to provide a basis for the moderation of democratic hedonism. To the extent that it demonstrates the natural status of our capacity for reciprocity and justice, it appears to provide a basis for restraining democracy's inclination to majority tyranny. To appreciate the plausibility of such claims, we must consider the problem posed by moral relativism and the proposed Darwinian refutation of it.

THE DARWINIAN REFUTATION OF RELATIVISM

Moral relativism, by its emancipation of bodily desire and self-interest, contributes to the democratic tendencies towards materialism and majority tyranny. When morality is widely held to be merely a matter of the individual's idiosyncratic choice or preference, moral principles become the mere servants of, and therefore incapable of restraining, passionately felt desires and interests. This is not to deny the theoretical possibility that one could both embrace moral relativism and then impose upon oneself, perhaps in an attempt at Nietzschean self-transcendence, moral rules contrary to one's desires and interests. These commitments imply no contradiction, since relativism means in theory only that any individual's morals may be whatever he wills them to be. Such uses of relativism, however, will be extremely rare in practice; and the predictable outcome of *popular* moral relativism will be the *popular* subordination of moral considerations to desire and self-interest. It is impossible to imagine masses of people *both* believing in relativism *and* willing to subordinate their desires to freely chosen conceptions of elevation of character or to subordinate their interests to freely chosen conceptions of the rights of minorities.

These implications of moral relativism for how men treat their own souls and their fellow citizens are not lost on contemporary Darwinian political theorists. Roger Masters contends that, when traditional grounds of morality have been rejected and the only basis of ethical judgment is taken to be "subjective value," it is "predictable from the theory of natural selection" that "most people" will "prefer their self-interest and pleasure." In practice, "relativism and nihilism" typically become "a rationalization for hedonism."[29] Similarly, Robert McShea notes that once the "secrets" of moral relativism —for example, that "all talk of value is mere confusion of speech" —"become open secrets, the general recognition of their truth produces the climate for the brazen practice of force and fraud at every level of human interaction."[30] While Tocqueville values religion as an authoritative denial of moral relativism, however, the Darwinian political theorists offer the insights of evolutionary biology as a no less forceful, and perhaps more persuasive, refutation of relativism.

More specifically, the Darwinian account suggests that the modern biological study of human nature undermines moral relativism by refuting its theoretical cousin and supporter, cultural relativism. Cultural relativism contends that there are no natural standards of right and wrong, that culture itself is the ultimate source of moral principles, and therefore that diverse cultures must be regarded as morally equal, since there are no trans-cultural standards of judgment. Such principles, however, destroy any solid reason for adherence to the morality taught by any particular culture. If all systems of value are morally equal because they are equally fictitious, equally products of human artifice, then it is not clear why one should adhere to any one instead of another. Thus cultural relativism throws the individual —or at least the individual who is aware of cultural differences—back upon his own idiosyncratic choice or preference. Again, however, the practical outcome of insisting on the mere conventionality of morality is morality's demise as a force capable of governing behavior. Once one is taught by cultural relativism that conceptions of nobility and justice are mere conventions that vary radically from culture to culture, such conceptions lose their ability to restrain our undeniably natural desires for pleasure and for the advancement of our own individual or group interests.

Contemporary evolutionary biology, however, appears to refute cultural relativism by demonstrating that morality is not merely conventional but in fact rooted in human nature. Defenders of cultural relativism commonly point to the great diversity of moral opinions as evidence that morality is not grounded in human nature. Nature, they point out, is universal, while convention is variable; and in the absence of universally recognized moral principles it seems that morality is based on cultural agreement rather than nature's command.

In response, the proponents of the new Darwinian political theory point out that contemporary evolutionary biology can explain such variability as consistent with a universal nature. "Contemporary biology," notes Roger Masters, "shows that the characteristics of an animal are not fixed and unaffected by its setting or life history."[31] Living nature, therefore, manifests considerable variability. For example, non-human animals of the same species—that is, animals that share a common nature—behave in diverse ways in diverse environmental circumstances. Larry Arnhart observes that "there is great variation from one chimpanzee group to another in tool use, botanical knowledge, and hunting techniques," such that "it becomes evident" that each group "has a social history that distinguishes it from others" and therefore that chimps are, like humans, "cultural and historical animals."[32] It would be ridiculous, however, to contend that this diversity demonstrates that the social behavior of chimpanzees is not rooted in their biological nature, that they are culturally self-creating beings. Rather, such variability points to the compatibility of nature and culture, to the notion that nature exists but is "not absolutely fixed," that it is modified in its expression by the influence of "developmental history" and "the external environment."[33]

Similarly, while humans admittedly display a great diversity of moral behavior and belief from one culture to the next, such variability need not be seen as evidence of the absence of a universal moral nature but of its modification by particular circumstances. To take but one example addressed by both Arnhart and James Q. Wilson, even the practice of infanticide—which, as an extreme manifestation of the variability of parental care of children, has been at times taken as evidence of "the purely conventional nature of family attachments"—actually suggests, when studied with sufficient attention, the natural status of feelings of parental love and obligation.[34] After all, if "mother-child attachments were purely a matter of convention," Wilson points out, one would expect to find infanticide practiced in a wide variety of social conditions, even where "food was plentiful, paternity certain, and the child healthy."[35] The evidence, however, does not bear out the cultural relativist presupposition. Indeed, Arnhart notes that "the anthropological record" indicates on the contrary that "[f]or most mothers killing a child is a painful action done only in harsh circumstances," usually when the child's paternity is uncertain or his survival doubtful because of poor health or lack of parental resources.[36] Thus infanticide, far from proving that the sense of parental obligation rests solely on culture, points instead to the conclusion that such a sense is indeed natural and can be overridden only by unusual and unfavorable circumstances.[37]

Modern biological science also suggests that this variability has been not only misinterpreted but overblown. Because some measure of cultural and moral diversity is compatible with the existence of a natural moral sense, the immediately visible variability that cultural relativists trumpet often masks more fundamental, universally present moral concerns. Thus Arnhart identifies twenty natural human desires —for such things as "parental care," "familial bonding," "friendship," and "justice as reciprocity," among others—that are "universally found in all human societies" and that "direct and limit the social variability of human beings as adapted to diverse ecological circumstances."[38] Similarly, Wilson finds behind the admitted diversity of moral rules among different cultures a universality of moral sentiments, such as "sympathy," "fairness," "self-control," and "duty."[39]

Cultural relativists have also sought to deny a natural basis for morality by pointing to the role of learning in moral development. Human beings, they contend, acquire the moral rules to which they adhere as adults through a process of parent-directed habituation while they are young. Surely what is natural emerges not by such an imposition from without but by a spontaneous generation from within. Thus they appeal to the opposition between what is innate and what is acquired, between nature and nurture. Modern evolutionary biology, however, indicates that this supposed opposition is in fact a false dichotomy, that in both animal behavior and human morality learned conduct specifies and is guided by natural propensities.

Contemporary biology, Arnhart suggests, points to "a complex conception of nature" as including "both original potential," or capacities that are present spontaneously, and "developed potential," or capacities that require learning to come to fruition. Thus, "[w]hile we commonly separate nature and nurture," science indicates that "ani-

mal nature . . . must be nurtured if it is to reach its natural completion." We find, then, that "monkeys and apes are such intensely social
animals that their natural development depends on social learning."
Chimps, for example, must "learn first from their mothers and then
from other members of their community an intricate repertoire of
social skills" in order to "satisfy their natural desires." Again, however,
no one would conclude on this basis that chimp sociability is purely
the result of learning, that it is not rooted in the animal's nature,
or that the chimp mind is, as Locke described the human mind, a
tabula rasa. Yet if such conclusions are unwarranted for other animals
they are unwarranted for men as well, and the previously mentioned
human desires that Arnhart notes may be regarded as natural even
though they develop only with the aid of "habit or learning."[40]

The common insistence on the opposition between nature and
nurture appears even less reasonable when we consider that, while
moral development clearly depends on learning, we appear to be
naturally predisposed to acquire the lessons of such moral nurturing.
This is suggested by how early and easily children begin to develop
a moral sense. Arnhart notes that even "within the first two years
of life infants show empathy, shame, and guilt, which provide the
emotional basis for a natural moral sense."[41] Similarly, Wilson points
out that "at an early age children can not only express notions of
preference, property, entitlement, and equity but can do so in such
ways that in fact alter the behavior of other children." Moreover,
their hold upon such ideas appears to grow stronger with the passage
of time, even as "parental control grows weaker," such that children
gradually come to discuss questions of conduct among themselves
"almost entirely without reference to adult authority or adult rules."[42]
Moral learning among human beings, then, like learning among other
animal species, is reasonably understood not as taking place apart
from nature, still less in opposition to nature, but instead as itself a
manifestation of nature.

Finally, on the Darwinian account the natural status of our moral impulses is revealed by their relationship to our natural interests. According to evolutionary biology, animal behavior, including the moral behavior of human beings, can be viewed as serving the animal's fundamental biological self-interest. As Francis Fukuyama points out, however, this self-interest is understood not in any such immediate terms as the individual animal's survival or pleasure but instead as the "passing on" of its "genes to offspring" and to succeeding generations.[43] Thus understood, the natural self-interest of individuals leads to what we commonly view as moral behavior through such mechanisms as "kinship" and "reciprocity." One has a certain genetic interest in the flourishing of one's kin, in their ability to survive and reproduce, because they share a certain portion of one's own genetic endowment. Thus individuals commonly sacrifice their own immediate pleasures or interests for the sake of the well-being of their nearest relatives. Moreover, since the benefits of mutual assistance can emerge under conditions of repeated interactions, one may ultimately gain even by assisting non-kin. Individuals will often forego immediate advantage in order to help another, in the hope of securing some reciprocal assistance in the future.

One might object here that the Darwinians end up destroying morality in their attempt to justify it, because their effort to ground it in nature subordinates it to biological self-interest. Surely by stressing the utility of cooperation and altruism, of morality and justice, they deny their status as moral principles at all. We tend to think that, in morality, intentions count as much as (and perhaps even more than) actions, and therefore that good acts done for the sake of self-advantage are not really moral at all. True morality must be pursued for its own sake, yet on the evolutionary view it merely serves genetic fitness.

The Darwinians have a response to this objection, and their theory can accommodate our sense that morality is good in itself.

Roger Masters contends that cooperative behavior among animals, including human beings, can usually be predicted on the basis of a "cost-benefit calculus." He immediately adds, however, that "[t]his calculus need not be conscious."[44] On the contrary, in many cases animals simply spontaneously behave, as a result of their evolutionary development, in ways, cooperative or competitive, that tend to secure their genetic interests. That is, they possess certain feelings evolved to advance their evolutionary interests, but they respond to those feelings not on the basis of any such calculation but simply because they experience the satisfaction of those feelings as good.

Such is the case with the cooperative and altruistic inclinations of human beings, the "moral sense" of which Wilson and the other Darwinian political theorists speak. As Fukuyama explains, because certain forms of moral behavior serve our genetic interests, over the course of evolution they tend to become encoded in our genetic makeup. Thus he suggests that "the people who are the actual products of this evolution have the cooperative tendencies embedded, so to speak, in their brain tissue," and that, for example, over time those who cooperate with non-kin "would enhance their own reproductive success" over non-cooperators "to the point where reciprocity" would be "coded into the genes governing social behavior." There is, then, a natural or biological basis for such sociable and moral feelings as "anger, pride, shame, and guilt."[45]

But this is as much as to say that we have an evolved capacity to experience moral behavior as pleasant or good in itself and immoral behavior as unpleasant. Fukuyama brings this implication to light when he points to evidence indicating that we have developed emotions to simplify our rational calculations of genetic self-interest. The "brain creates numerous somatic markers"—"feelings of emotional attraction or repulsion"—that provide the basis for our experiencing moral activity "as an end in itself" and "heavily invested with emotion."[46]

Put another way, what we call moral behavior emerged because those who practiced it were more likely to succeed in reproducing their genes in future generations. But those who experience such behavior as pleasant, and its opposite as unpleasant, are much more likely to practice it. As Robert McShea puts it, the "feelings of humans and of the other higher animals came into being under the pressure of evolutionary selection, but they are presently experienced as simple imperatives," as "what they are, not as passions for survival."[47] Just as sexual pleasure serves genetic fitness but is experienced as desirable for its own sake, so certain moral passions, like mother-love or a sense of reciprocal fairness, serve genetic fitness but are experienced as satisfying in themselves. Hence we may say that natural selection has favored the formation of conscience—of a moral sense—understood as a set of feelings that prompt us to cooperative action and deter us from its opposite.

CONCLUSION

Early in *The Moral Sense*, James Q. Wilson relates that his study of crime led him to think, paradoxically, that "[w]hat most needed explanation" was "not why some people are criminals but why most people are not." Given the instinctive status of our selfish inclinations, our desires for "food, sex, wealth, and self-preservation," it might seem "puzzling" that they should so often be held in check by apparently learned traits like "self-control, sympathy," and "a sense of fairness." The answer to this riddle, he argues, is that, as evolutionary biology suggests, these principles of restraint, these moral inclinations, "are sensibilities whose acquisition is as much a product of our human nature as the appetites they are meant to control."[48] Human nature includes "a desire not only for praise but for praiseworthiness, for fair dealings as well as for good deals, for honor as well as advantage."[49]

The Darwinian refutation of relativism, then, offers an apparent solution to the democratic problems that Tocqueville identifies. By showing that the desire for moral self-esteem is no less natural than the desire for material ease, and that concern with justice is no less natural than self-interest, the Darwinian account of human nature seems well-suited to moderate the democratic inclination toward materialism and majority tyranny. Those who succumb to such vices, the Darwinian can point out, do violence to their own nature and thereby diminish their own happiness. There are perfectly intelligible natural reasons for democratic citizens to cultivate both respect for themselves and for the rights of others, and we have no need to appeal to the moral teaching of revealed religion.

Ultimately, however, the teaching of the Darwinian political theorists fails to fulfill its promise. To see why, we need to consider more closely democracy's moral needs and the substance of the morality that can be based on evolutionary grounds.

⁓ 3 ⁓

Ennobling Democracy

WHATEVER ITS MERITS ON ITS OWN TERMS (an issue to which we will return in chapter 6), the Darwinian refutation of relativism does not in the end suffice to moderate democratic materialism in the way that Tocqueville hopes religion will. That materialism, as we have seen, is not amoral. On the contrary, it is accompanied by respect for public order and therefore recognizes and adheres to the distinction between licit and illicit pleasures. Democracy requires, then, not merely a justification of morality as natural and thus on a par with our natural hedonistic appetites, but a justification of a lofty morality that transcends the petty materialism of ordinary democratic life. Or, as Tocqueville suggests, democracy needs, as a supplement to its decent but low public morality, a morality that fosters human greatness. While the evolutionary account of human nature seems to promise, and its proponents seem to desire, such a morality, the attempt to ground principles of right in the Darwinian view of nature finally sustains only a morality indistinguishable from the decent materialism that democracy needs to transcend.

DARWINISM'S TRUNCATED ARISTOTELIANISM

The new Darwinian political theory appears initially to offer a scientifically justified recovery of classical political philosophy's lofty concern with virtue understood as moral excellence. Modern political philosophy presents itself as a repudiation of the thought of the classics, a more realistic account of human nature that recognizes man's natural selfishness and indifference to virtue. Contemporary evolutionary biology, however, provides a scientific refutation of the modern account. While thinkers such as Thomas Hobbes and John Locke contend that morality and sociability arise only as artificial institutions devised by men to serve their natural interest in their comfortable self-preservation, contemporary Darwinian theory indicates that, in the words of Francis Fukuyama, "human beings are *by nature* social creatures, whose most basic drives and instincts lead them to create moral rules that bind themselves together into communities" and whose natural rationality "allows them to create ways of cooperating with each other spontaneously."[1] Thus man's natural state is not, as Hobbes argued, a war of all against all—from which we escape only through a self-interested calculation of the need for an all powerful state that can impose order—but instead "a civil society made orderly by the presence of a host of moral rules."[2] In repudiating the modern theorists of comfortable self-preservation, the Darwinian account of human nature seems to point with approval to the moral wisdom of the ancients and therefore to promise a morality that will transcend the materialism of democratic modernity, of which such thinkers as Hobbes and Locke are arguably the intellectual architects.

This impression is strengthened by the Darwinians' claim, made explicitly and repeatedly, that the contemporary life sciences restore the credibility of Aristotle's understanding of man as a naturally moral and political animal. For example, Fukuyama claims that modern "evolutionary biology" can "wholeheartedly agree" with Aristotle's presentation of humans as political animals, naturally organizing themselves into political communities "whose existence is necessary for the complete satisfaction of what humans by nature desire" and who are "capable of the moral virtues necessary to sustain such communities."[3] Larry Arnhart observes that Darwin agreed with Aristotle in claiming "that human beings are by nature social animals," uniting "first in families and then in larger social communities," and "in deriving morality from human nature."[4] James Q. Wilson suggests that, in providing evidence for the natural status of the moral sense, "[s]cience supplies more support for the 'ancient' view of human nature than is commonly recognized."[5] Finally, Roger Masters contends that his "naturalistic approach" to the study of human society, based on "contemporary research in the biological sciences," directs us "to a view of 'natural justice'" and thus returns us to "views akin to those of Aristotle."[6]

Aristotle's moral and political teaching is undoubtedly lofty. The purpose of politics, and indeed a primary purpose of human life, he famously argues, is to foster moral excellence, understood as the activity of the soul in accordance with certain difficult virtues. These include courage, by which one masters the fear of death in fighting for a worthy cause, temperance, by which one consents only to such bodily pleasures as right principle can approve, and liberality, by which one gives freely to those who are deserving. We pursue these virtues for the sake of their own nobility or beauty. By their precedence over the goods of the body, by the evident discipline they impose on bodily desires, and by Aristotle's requirement that they not be sought with any ulterior, hedonistic motive, such virtues transcend the decent materialism of modern democracy. Any theory that could establish the scientific basis of Aristotelian morality would clearly answer one of the primary purposes of religion in Tocqueville's political science.

Despite this apparently promising beginning, however, the Darwinian account soon reveals serious substantive differences with the Aristotelian understanding of human nature and morality. The possibility of such differences is first suggested by the Darwinian political theorists' conviction that the modern, scientific study of human nature generates knowledge more reliable or complete than the fruits of the more traditional and less sophisticated methods of previous philosophers. Roger Masters, for example, remarks that in the study of human nature we have "[p]ast philosophers" at a "disadvantage." We enjoy a "theoretical understanding of the mechanisms of evolution that is more precise and complete than at any prior epoch." And we have access to "data," such as "fossil evidence" and "comparison of the genetics and biochemistry of different species," far better than that "available even fifty years ago." Not surprisingly, then, Masters concludes not that evolutionary biology merely confirms the wisdom of the ancients but that it plays a "central role in deepening our understanding of the human condition."[7] Similarly, Fukuyama holds that while modern evolutionary science confirms the general Aristotelian view of humans as naturally moral and sociable, it also "allows us to be much more precise" than Aristotle "about the nature of human sociability," about "what is and is not rooted in the human genome."[8]

A hint of the differences between Darwinian and Aristotelian morality is provided by Fukuyama's comment that his evolutionary argument establishes that "[b]oth nature and rationality" support the development of such traits as "honesty, reliability, and reciprocity," or what he sums up as "the ordinary virtues."[9] Such a formulation would be absurd to Aristotle, given his repeated characterizations of virtue as a kind of excellence or nobility. James Q. Wilson betrays a similarly non-Aristotelian spirit when he refers to those of good character as, in "common parlance 'nice persons' or 'good guys,' or in polite (and vanishing) discourse, 'ladies' and gentlemen.'"[10] The term "gentleman," of course, is regularly employed to translate an expression Aristotle uses with some frequency: *ho kaloskagathos.* Literally, however, this term means "the noble and good man," who seems to stand at some considerable distance from the "good guy." The former is excellent, the latter merely decent or sociable. Thus the Darwinian political theorists tend to reduce morality to mere sociability or decency in a way that is alien to Aristotle's thought.

Throughout *The Great Disruption*, Fukuyama equates virtue or morality with a kind of mutually beneficial cooperativeness that tends to support an orderly and decent society, one in which the individual's natural interest in survival and reproduction can readily be attained. He notes that altruism and reciprocity, the foundations of moral behavior, tend to be in the self-interest of individuals and, indeed, asserts that "almost all behavior we understand to be moral involves two-way exchange of some sort and ultimately confers mutual benefits on the parties participating in it."[11] Although he acknowledges the importance of the Kantian emphasis on purity of intention, and thus pays some tribute to the similar Aristotelian notion that the virtues must be practiced for their own sake, he emphasizes much more the notion of mutually beneficial "moral exchange." Such a concept is certainly not alien to Aristotle, who devotes considerable attention to justice, which involves mutually advantageous reciprocity. In Fukuyama's account, however, "moral exchange" seems to exhaust the meaning of natural morality, which is clearly not the case for Aristotle.

One of Fukuyama's primary concerns is to argue that, given the natural status of our cooperative tendencies, the maintenance of a decent and orderly society is virtually guaranteed, at least in the long run. Hence his contention that Burke was wrong in holding that the Enlightenment's destruction of traditional religious and moral authority would in the end cause society to "implode."[12] Here again, his argument betrays a decidedly non-Aristotelian spirit, for he fails to address Burke's eminently Aristotelian concern, not so much that Enlightenment rationalism would cause society to fall apart, but that it would deprive it of its elegance or generosity, its moral beauty.

Fukuyama's defense of capitalism against those who claim that it simply undermines morality also reveals that he does not share the full range of Aristotle's moral and political concerns. While admitting that the economic and social dynamism of capitalism does in fact undermine traditional moral norms, Fukuyama contends that emphasis on this alone is "one-sided" since capitalism also tends to create new norms. In fact, it probably "is a net creator of norms and thus a net moralizing force in modern societies."[13] Indeed, given the natural status of our cooperative inclinations, one would expect that capitalism's system of free interactions would tend to give rise to moral behavior, at least as Fukuyama has characterized it. He quotes approvingly from Samuel Ricard that "commerce" causes a man to learn "to deliberate, to be honest, to acquire manners, to be prudent and reserved in both talk and action. Sensing the necessity to be wise and honest in order to succeed, he flees vice, or at least his demeanor exhibits decency and seriousness so as not to arouse any adverse judgment on the part of present and future acquaintances."[14]

This assertion that capitalism is a net creator of norms raises the question, "What kind of norms?" Fukuyama's quotation from Ricard immediately indicates the answer: it creates the kind of norms that allow one to get along well in society of self-interested people, though not necessarily those that would foster nobility or beauty of character. His distance from Aristotle is similarly suggested in his use of Montesquieu to defend the morality of capitalism. Fukuyama quotes the famous observation, contained in *The Spirit of the Laws*, that "commerce . . . polishes and softens barbaric ways."[15] His ellipses, however, omit Montesquieu's other observation, that commerce also "corrupts pure mores, and this was the subject of Plato's complaints."[16] No doubt Aristotle's view of commerce is less severe than Plato's, but it would still be far from easy to enlist him among the supporters of modern capitalism. For though he does famously offer a defense of private property in Book II of the *Politics*, Book I contains a rather flat condemnation, as contrary to nature, of the spirit of unlimited acquisition.

This reduction of morality to decent sociability is equally evident in the arguments of the other Darwinian political theorists. Larry Arnhart, for example, notes that, for Darwin, morality arises from the "feeling of pleasure from society," which is in turn "probably an extension of the parental or filial affections." "[O]ur moral sense . . . originat[es] in the social instincts" and is "largely guided by the approbation of our fellow men."[17] In a similar spirit, Wilson asserts that "[o]ur moral nature grows directly out of our social nature" and that the "mechanism underlying human moral conduct is the desire for attachment or affiliation." Thus, "to live life on human terms" is to "acknowledge the obligation of one's social nature by discharging duties to family, friends, and employer, duties that rest on reciprocal affection, interdependent needs, the bonds of sympathy, and the requirements of fair play."[18]

These are admittedly worthy aspirations, but they hardly exhaust Aristotle's understanding of what it means to live a human life. Indeed, while the Darwinian account presents virtue as aiming to sustain a decent society that serves our most commonplace desires, Aristotle seems to suggest nearly the opposite: that the political community ultimately exists to allow the occasion for the display of virtues that are noble or beautiful in their own right. To be sure, Aristotle acknowledges the origins of the political community in the same interests that the Darwinian account indicates are the source of our moral and sociable nature: preservation and reproduction. And, to the extent that he thus includes the "striving to leave behind another that is like oneself" along with the desire for self-preservation as the natural basis of human association, Aristotle may fairly be placed with Darwin in opposition to moderns such as Hobbes, who makes self-preservation the sole principle of his political teaching.[19] Aristotle almost immediately suggests, however, that human community, and hence human morality, transcend these common concerns. The city, he says, "while coming into being for the sake of living," "exists of the sake of living well."[20] That this mode of living well transcends living according to the ordinary strivings that first gave rise to the city is made explicit later, when Aristotle asserts that the "political partnership must be regarded" as "being for the sake of noble actions, not for the sake of living together."[21] Morality, it seems, transcends sociability.

Turning to Aristotle's account of virtue in the *Nicomachean Ethics*, we again find many indications that morality cannot be reduced to the service of such ordinary aims as survival and reproduction, or understood merely as an outgrowth of the natural moral and sociable feelings that support their achievement. For moral nobility is repeatedly presented by Aristotle as being in tension with those aims and feelings and the society to which they give rise. Consider Aristotle's argument in Book II of the *Ethics*. He contends that virtue is concerned with pleasure and pain, in part because they frequently cause us to do what is base and to refrain from what is noble. He subsequently adds, however, that pleasure and pain arise from the passions, and he includes among the passions such evidently sociable inclinations as "friendship" and "pity."[22] Hence such sociable and (in the ordinary or low sense, the Darwinian sense) moral inclinations, far from being simply the source of nobility, can in fact be an impediment to it.

Aristotle does give the emotional bases of sociability and decency their proper due. In Book III of the *Ethics* he implies that we have a natural fear of bad reputation and lack of friends, and hence a natural desire for good reputation and friendship. He suggests that such fears are good and even noble, since to be without them is to be shameless and hence capable of the worst deeds.[23] Nevertheless, his account also indicates that such desires and fears are good only in a limited sense, because they must be disciplined in order to achieve a greater natural good, the happiness that arises from noble action.

True virtue will require us to temper our ordinary sociable and moral inclinations, and hence it may not always be in harmony with the goods those inclinations seek to achieve and will not allow us simply to be successful or make our way well in society. Such a possibility is also indicated by the fact that, according to Aristotle, it is possible, even commonplace, for the virtuous man to be mistaken for something else. Every virtue, Aristotle famously teaches, is a mean between two vicious extremes. But in each case the mean appears vicious to those at either extreme. Hence "a coward calls a brave man rash and a rash man calls him a coward, and correspondingly in other cases."[24] Aristotle subsequently suggests, however, that human nature is more inclined toward one of the extremes in each case.[25] Thus it seems that the display of any virtue will cause one to be viewed as vicious by most people. Again we see the need to discipline one's ordinary sociability and morality, one's concern with one's good standing in society, in order to practice virtue in the noble, Aristotelian sense.

Aristotle notes along these lines that because most men are excessively fond of "fun and raillery" buffoons are thought to be "witty and pass for clever fellows."[26] Presumably, the witty, because they fall short in giving the amount of pleasure expected by most, are blamed as boors. One can discern the same difficulty in the case of each of the virtues Aristotle discusses in the *Ethics*. Most men are excessively desirous of receiving benefits, and therefore the truly liberal man, who will give only to worthy causes, will be viewed by the many as stingy. Similarly, since the virtuous man's moral standards will no doubt far surpass those of most people, in displaying true friendliness by consenting only to what is fitting, and opposing what is not, he will be seen by most as surly or quarrelsome.

Aristotelian nobility's independence of and tension with the ordinary ends of society and the feelings supporting them, and hence its distance from evolutionary morality, is perhaps best illustrated with reference to Aristotle's account of courage in the *Nicomachean Ethics*. Larry Arnhart points out that one can devise a plausible evolutionary account of the origins of a kind of courage. This seems at first unlikely, since "courageous individuals naturally inclined to sacrifice their lives in defense of their community might often leave fewer offspring than cowardly individuals." Nevertheless, as Darwin pointed out, "in the competition between groups those with the more courageous members would often prevail," and therefore "[n]ot only courage but other moral dispositions to social cooperation might strengthen one group against other."[27] This argument seems to imply that courage simply serves the well-being of the community, and Arnhart makes this clear elsewhere. "In evolutionary history," he remarks, "courage meant that individuals would fight for the interests of their tribal group against other groups."[28]

In harmony with his general approach, which tends to assimilate Darwinism and Aristotelianism, Arnhart presents the Darwinian conception of this virtue as consistent with Aristotle's. He points to Aristotle's depiction, in the *Nicomachean Ethics*, of "the manly courage of the warrior-citizen facing death in battle," which "epitomizes virtue as virility in asserting one's group against competing groups."[29] Nevertheless, courage as Aristotle presents it cannot be reduced merely to a disposition to cooperation with a view to the defense of the community and cannot be understood to arise simply from our sociable and cooperative inclinations. Aristotle does speak of a kind of "citizen's courage," which bears a strong resemblance to that described by Arnhart. Those who possess this quality, Aristotle suggests, face fear in battle because of the desire for honor and the fear of dishonor, and will "sacrifice their lives in defense of their community," often proving themselves better soliders than professional mercenaries.[30]

While the Darwinian account seems so far to agree with Aristotle's, the radical gulf between them comes into view on a more careful consideration of Aristotle's position. Aristotle does not regard citizen's courage, which seems to arise from our sociable concern with the interests of our group and the esteem of its members, as true courage, which, again, must be pursued only with a view to its own nobility or beauty. On Aristotle's understanding, the virtue of the truly courageous man is much more problematic in its relationship to the well-being of the community and in its responsiveness to the community's honorable inducements to martial service. Those who possess all virtue, Aristotle contends in this context, may not make the best soldiers. Their lives are happy, and they are therefore reluctant to put them at risk. And such men cannot be induced to fight by the desire to gain honor by serving the city but only by the prospect of achieving what is noble.[31] Aristotelian courage, like Aristotelian virtue generally, does not, unlike its Darwinian analogue, arise simply from our natural sociability and seems in fact to transcend it.

In sum, one may formulate the differences between Aristotle and Darwinism as follows. Virtue as the Darwinian political theorists understand it is the set of sociable and cooperative inclinations that support our living together in an orderly and decent society. Virtue as Aristotle understands it requires that we control the sociable and cooperative inclinations that serve such ends so that we might achieve a higher end and fulfill a higher natural desire. One might put it as starkly as this: virtue as Aristotle understands it requires that we transcend virtue as Darwinism understands it.

In demoting morality to the service of common desires, the new Darwinian political theory veers back in the direction of the

modernity it claims to repudiate. This tendency is most obvious is
Fukuyama's *The Great Disruption*, in which his apparent criticism
of the moderns rings a bit hollow in light of his embrace of the
very regime to which their thought ultimately gave rise. Fukuyama
endorses liberal democracy, says that "societies created on . . . individualistic premises have worked extraordinarily well," and claims
that "[i]ndividual self-interest is a lower but more stable ground than
virtue on which to base society."[32]

As his book nears its end, he reassures the reader that we may
expect a spontaneous "renorming" to occur based on our natural
moral proclivities. Fukuyama, however, also notes that that renorming
will likely go much further in the realms of crime and trust than in
sexual activity and family relations. This is so, he continues, because
"strict rules about sex make sense in a society in which unregulated
sex has a high probability of leading to pregnancy, and where having
a child out of wedlock is likely to lead to destitution, if not early
death, for both mother and child."[33] On Fukuyama's Darwinian understanding, it seems, moral rules "make sense" not for promoting
a noble self-control but for preventing material harm. Here one is
reminded of Hobbes's suggestion that moderation or temperance, as
one of the private virtues, aims merely to prevent "the destruction"
of individuals.[34]

In the end, the Darwinian account of morality, while rejecting
Hobbes and embracing Aristotle in one sense, rejects the latter and
embraces the former in a perhaps more fundamental sense. Darwinism agrees with Aristotle that morality is naturally rooted in
our spontaneous desires and therefore denies the Hobbesian (and
generally modern) assertion that the virtues are mere artificial cre-
ations or "theorems found out by reason." On the other hand, the
weight of the Darwinian argument suggests that the Darwinians
would agree with Hobbes's typically modern assertion that the moral
virtues are finally no more than "the meanes of peaceable, sociable,

and comfortable living," and with Hobbes's accompanying criticism of the previous "Writers of Morall Philosophie"—and he surely had Aristotle in mind—who failed to see this.[35] Virtue understood in this sense, however, is compatible with democratic materialism and, as Tocqueville's argument suggests, ordinarily respected by democratic men even as they single-mindedly pursue physical comforts. The Darwinian account of human nature, then, while perhaps usefully demonstrating the natural basis of a limited kind of morality, seems to offer nothing in response to Tocqueville's concern with elevating the moral tone of democracy.

DARWINIAN NATURALISM AND MORAL MEDIOCRITY

The Darwinian political theorists and Aristotle both seek to ground human morality in nature, yet their understandings of morality are very different: the latter's is noble while the formers' is merely decent. The cause of this moral difference is the very different understandings of nature to which these two approaches appeal. While Aristotle seeks nature in what is perfected and understands such nature ultimately with a view to the transcendent, Darwinism seeks nature in the merely universal or common and understands such nature to arise from a decidedly worldly concern: the evolutionary interest in survival and reproduction. Because of its understanding of nature, Darwinism cannot support a morality that aspires to a generous disregard for genetic self-interest, or even a principled subordination of it to loftier concerns. The result is, as we have seen, that when Darwinism and Aristotle both say that human morality is natural, their agreement is far more formal than substantive.

The Darwinian political theorists' emphasis on the natural as the universal, the typical, or the common is most clearly evident in the work of Arnhart and Robert McShea. "Human nature thinkers," McShea observes, "are concerned with the study of the feeling pattern typical of our species" and their "value theories" are accordingly "based on their understanding of the normal range of human feelings."[36] Arnhart similarly bases his moral understanding on "the species-typical pattern of moral sentiments," and generally throughout his book associates the universal with the natural and, hence, the good.[37] For example, he rejects as "imprudent dogmatism" the "Catholic Church's prohibition on divorce," since it "is contrary to the natural pattern of human mating," which most commonly manifests itself as "serial monogamy."[38] On the view implied by this criticism, it seems, the good equals the natural, which in turn simply equals what usually happens.

This is not to say, of course, that it is wrong to seek clues to human nature in universally manifested human behaviors. Rather, it is to suggest that because it derives its ethics from an account of nature limited to the universal or common, without making reference to any other principle, the Darwinian understanding cannot rise above moral mediocrity. The Darwinians recognize psychopathy, the absence or pathological weakness of sociable and moral feelings, as in some sense a natural phenomenon, since it occurs in nature and has natural causes, either genetic or environmental. They conclude, however, that psychopathy is not natural in the sense of normative: the psychopath, in his incapacity to experience sociable emotions, deviates widely from the species-typical feeling pattern, and therefore his behavior cannot be considered desirable for ordinary human beings.[39] No doubt this is true. The difficulty, however, is that, in the absence of any moral principle besides the normal or universal, the Darwinian account would have to take the same position with regard to other human beings who manifest unusual feelings and behaviors, such as the philosopher, the saint, and Aristotle's *kaloskagathos*, or gentleman, whose noble conduct deviates sufficiently from that of ordinary and less-than-virtuous human beings that his life is in constant tension with the society in which he finds himself. Such human beings, no less than the psychopath, must be viewed as morally irrelevant by the Darwinian account: they are not exemplars but mere outliers on the bell curve of human nature.

The evolutionary account of the origins of morality in our natural interest in genetic fitness also tends to prevent Darwinian virtue from manifesting itself in any very lofty or generous form. On the Darwinian view, the fundamental interest of any organism lies in getting its genes reproduced in the next and in succeeding generations. Such genetic selfishness, however, can give rise to moral and cooperative behavior among organisms through kinship and reciprocity: since kin share a portion of one's genes, one has an interest in their reproductive success as well as one's own, and since cooperation even with non-kin can enhance one's chances of successful reproduction, assistance is to be rendered when reciprocity can reasonably be expected. As noted in chapter 2, because of natural selection's ability to generate spontaneous feelings that generally support genetic fitness, such altruistic or cooperative behavior need not be experienced by the organisms engaged in them as mere self-interested strategies.

Nevertheless, even though prompted by directly felt desires to aid one's kin or engage in mutually beneficial cooperation, such moral behavior will not, on the evolutionary account, range far beyond the genetic self-interest that gave rise to it in the first place. Thus Roger Masters notes that, although cooperative behavior among organisms need not be based on a "conscious" weighing of genetic interests, such behavior will nonetheless "vary as if it were a strategy chosen on the basis of cost-benefit optimization."[40] The evolutionary account, then, gives us little reason to expect, or even to desire, to see human beings acting morally toward those to whom they are not related or with whom they are not accustomed to engaging in reciprocal cooperation. Accordingly, Arnhart dismisses the notion that we "can invoke ethical norms, with no grounding in empirical reality, which transform human beings into selfless creatures who act for the benefit of others with no compensating benefit for themselves" or for those to whom they are "attached by various social bonds."[41] Of course, care for kin and reciprocal cooperation are perfectly respectable activities and are necessary to human flourishing. But the point here is that such principles do not seem to rise above ordinary decency, as Aristotle's principles do, and as Tocqueville's moral correction of democracy requires.

One might note in addition that even within its limited sphere of kin and non-kin cooperators Darwinian morality does not stray far from the genetic interests of the individual animal. Reciprocity, after all, boils down to "tit-for-tat," or rendering aid to whose who render aid in return.[42] This, Masters suggests, "is more reasonable on naturalistic grounds than an a priori ethics ignoring the past behaviors of others."[43] One assumes, then, that merely sharing food with someone who is hungry is unreasonable in comparison to sharing food with someone who you may reasonably expect will share with you in the future. Presumably, similar self-interested limits would be at work even in one's relationship to kin. Because aid to those who share one's genes can serve genetic fitness even in the absence of reciprocity, one would no doubt be more inclined to outright altruism with regard to kin. Nevertheless, the common interest arising from kinship would not result in a principled commitment to even a close kinsman's interests in preference to one's own. For example, on the evolutionary understanding, at some point the well-being of a handicapped child would come up against the parents' realization of their ability to generate new and less demanding offspring.

NATURE, RELIGION, AND THE FULL STATURE OF MAN

In contrast to the Darwinian political theorists, Aristotle adopts a different understanding of nature. While consulting the universal as a clue to the natural, and while recognizing the rootedness of some portion of human nature and human morality in the spontaneous desires arising in the service of our biological interests in survival and reproduction, Aristotle's understanding of nature reaches further, finally embracing notions of perfection and of the transcendent. Hence his morality's ability to rise above the mediocrity of commonly observed behavior and the narrowness of ordinary sociability in the service of genetic fitness.

The gap between the evolutionary and Aristotelian accounts of nature, and hence of morality, can be well illustrated by an examination of their differing treatments of the political significance of religion. As we have seen, the Darwinian political theorists downplay the political and moral importance of revealed religion. Fukuyama, for example, comments that the "ordinary morality" to which he assigns a natural status will come about "*spontaneously*" and "*without the benefit of a prophet who will bring the word of God.*"[44] In contrast, Aristotle includes what appears to be revealed—or what Fukuyama terms "hierarchical"—religion among the functions needed for the operation of the best regime, which aims at the fulfillment of human nature. Indeed, he hints at such religion's centrality to that aim, insofar as among the things necessary for the best city he lists "fifth, *and first*, the superintendence connected with the divine, which they call priestcraft."[45]

This difference would seem to arise from the relative loftiness and difficulty of the kind of virtue that Aristotle's best city seeks to foster. It is, again, a kind of nobility of action that rises above the spontaneous sociability and cooperation to which Fukuyama and the other Darwinian political theorists give the name virtue. Virtue as they understand it, being rooted in our most common natural inclinations, does not need the support of revealed religion. Fukuyama indicates that our natural sociability and morality is such that children separated from their parents and community will spontaneously create a moral order not much different from that they have left behind.[46]

This raises the following question: If, as Aristotle insists, nobility is rooted in our nature, if we have a natural directedness toward it, why does it need the support of revealed religion? After all, Aristotle holds in a famous passage that virtue is pleasant by nature.[47]

The same passage suggests, however, that most people do not find virtue pleasant; and of course much of Aristotle's teaching indicates that while nobility is pleasant by nature we do not immediately or spontaneously experience it as such. The *Ethics* concludes by noting the need for a politically supervised habituation to "secure that the character shall have at the outset a natural affinity for virtue, loving what is noble and hating what is base."[48] Nobility may be somehow the core of our nature, but it is not at first the most powerfully felt part of our nature. Nor is this surprising, in light of Aristotle's teaching that a thing's nature is to be understood as its completion or perfection and not merely as its origins.[49] Virtue therefore requires political education, apparently including that provided by publicly sanctioned revealed religion.

This dispute over the public utility of religion, however, points to a more fundamental difference over the standing of religion in nature. Aristotle, it seems, affirms for religion a kind of natural status, insofar as he places it among the necessary equipment of the best regime. But going further, we might say that Aristotle comes close to acknowledging something like a natural religious longing. Aristotle is clearly not a religious thinker, and in fact his discussion of the divine in Book x of the *Ethics*, as well as in the *Metaphysics*, calls into question, to put it mildly, whether the divine is cognizant of human affairs. Nevertheless, Aristotle's presentation of our natural directedness toward the noble can be viewed as quasi-religious insofar as it appears as a kind of longing for the transcendent. Aristotle indicates in Book x of the *Ethics* that the noble, which is said throughout the *Ethics* to be the aim of moral activity, is one of the objects of intellect or *nous*, which is itself somehow divine and through which one can achieve a kind of participation in the eternal.[50] But if *nous* participates in the divine and eternal, so, it would seem, do its objects. Hence our natural desire for the noble is a natural transcendent longing.

Aristotle's understanding of human nature as including a longing for a kind of supernatural perfection roughly corresponds to Tocqueville's view. While *Democracy in America* nowhere offers a systematic definition of human nature, it is nonetheless evident that Tocqueville's evaluation of democracy is made in light of some understanding of the nature of man. Tocqueville's account suggests, however, that human nature includes an element that is attracted to the infinite and that is therefore the basis of a greatness to which other animals cannot aspire. He remarks that while "the human mind leans at one extreme toward the bounded, material, and useful, at the other it naturally rises toward the infinite, immaterial, and beautiful."[51]

This natural inclination toward the infinite or transcendent manifests itself in a number of ways. First, Tocqueville indicates that it can be seen in the human desire for wisdom for its own sake, the impulse to pursue theoretical science or philosophy, as distinct from merely useful knowledge. He refers to "an instinctive penchant" that tends to "elevate" the human mind "toward the highest spheres of the intellect," which seem to involve a concern with "the most theoretical principles, the most abstract notions, those whose application is not known or is very distant."[52] This transcendent longing also reveals itself in an impulse toward a generous and self-forgetting morality. Thus Tocqueville suggests that human beings aid each other not only out of enlightened self-interest but also by "abandoning themselves to the disinterested and unreflective sparks that are natural to man."[53] Finally, Tocqueville indicates that this transcendent longing manifests itself as a directly religious impulse. He contends, for example, that this life and its "incomplete joys" cannot "confine the whole imagination of man" or be sufficient for "his heart." "Alone among all the beings," he continues, "man shows a natural disgust for existence and an immense desire to exist," and these conflicting "instincts constantly drive his soul toward contemplation of another world," to which we are directed by "religion." Tocqueville concludes that religion draws its strength from one of the "constituent principles of human nature." [54]

As these passages indicate, Tocqueville may emphasize more than Aristotle the spontaneous power of man's longing for the eternal. Nevertheless, like Aristotle he accords revealed religion an important role in supporting such desires. Under modern, democratic conditions, Tocqueville's account suggests, the human soul is powerfully distracted from such lofty concerns and toward the mundane and ordinary. This explains Tocqueville's emphasis on religion's role in fostering the human greatness founded on the soul's loftier impulses.

In contrast to the Aristotelian and Tocquevillian accounts, the evolutionary understanding of human nature tends to deny the natural status of the human directedness toward the transcendent. Fukuyama, for example, implicitly affirms the artificiality of such longings when he suggests that what he calls "folk religion"—or beliefs that are merely supportive of cooperative community norms—alone is of natural origins. He suggests that the "renorming" that is taking place will likely include a religious revival, but not in the sense of a revitalization of revealed religions that make demanding claims. Instead, people will turn to decentralized forms of religion, ones that do not press any ultimate truth claims, simply as a way to affirm "the community's existing norms and desire for order." Such "religion" is fundamentally this-worldly in its outlook, and such religion alone, on Fukuyama's account, can be said to be rooted in man's nature. This becomes explicit when Fukuyama suggests that religion is merely "a reasonable extension of the natural desire for social relatedness that all human beings are born with"—and not, as Tocqueville would have it, a manifestation of our longing for a transcendent and eternal perfection. [55] Larry Arnhart also appears to deny that humans have a natural aspiration for the eternal. He suggests that the "transcendent longing" or "need for redemption from the world" that many human beings display is merely "acquired," thus implicitly denying such longings any place in the natural constitution of the soul.[56]

Ultimately, then, Darwinism, on the one hand, and Aristotle and Tocqueville, on the other, offer moralities with markedly different horizons. The Darwinian looks only to the realization of our most immediately experienced natural inclinations, the desires for life and reproduction, and the sociable and moral feelings that support them. Aristotle and Tocqueville, without denying the natural status of these inclinations, look beyond them to the realization of a transcendent longing for the noble and divine that is not only also a part of human nature, but in fact its center and completion. When Fukuyama, and the Darwinians generally, assert the agreement of contemporary evolutionary biology with Aristotle's contention that political community "is necessary for the complete satisfaction of what humans by nature desire," that agreement seems finally to be more apparent than real. Despite the overlap in their views of natural human inclinations, in the end they do not fully agree on which desires are natural, much less on which desires are at the core of our being.

Put another way, from Aristotle and Tocqueville's standpoint the evolutionary account of human nature offers us only a truncated man. In fact, one might say that in the Darwinian account, man as Aristotle and Tocqueville present him disappears. In the absence of a natural desire for the eternal and noble, which Aristotle suggests is what distinguishes man from the other animals,[57] man appears merely as a more sophisticated version of other creatures, differing from them in no fundamental natural way. The Darwinian account of human nature implies as much when it tries to explain man's entire sociable and moral nature on the basis of biological interests shared with all living things.

DARWINISM'S FAILED NOBILITY

One might respond that this Darwinian diminution of man is only a defect if one embraces the loftier view of human nature advanced by Aristotle and Tocqueville. But perhaps one need not embrace it. Perhaps the elevated and disinterested virtues of which Aristotle and Tocqueville speak, as well as the supposedly natural human desire for the eternal of which they are said to be manifestations, are simply illusory. On this view, the evolutionary account of human nature does not wrongly diminish man's stature but merely sees it as it really is.

The reader might well question whether such a defense is acceptable, in light of its implication that we need not, as individuals or as a community, aspire to anything beyond orderly hedonism or decent materialism. In any case, the Darwinian political theorists seem to deny themselves the luxury of this defense. For despite the tendencies of their evolutionary understanding of human nature, the Darwinian political theorists evidently do not want to diminish man's stature in this way, nor are they unmoved by the attractions of a noble morality. On the contrary, they all seem to come to a point at which they affirm the importance of more elevated virtues and try to justify them on the basis of evolutionary naturalism. The efforts fail, however, and the paradoxical result is that the Darwinian political theorists themselves bear inadvertent witness to the unacceptable narrowness of Darwinism's account of human nature.

Among these theorists, James Q. Wilson appears to be the least interested in the virtues that transcend the ordinary decency supported by our evolved desires. Yet he does speak with admiration of such virtues as "courage" and "generosity," which would seem to go beyond the mere sociability with which his account of morality is primarily concerned.[58] At one point, Wilson notes that we "praise people for being generous," "praise them most when their generosity is uncompensated," and "reserve our highest praise for heroes who sacrifice their lives that others may live."[59] Such behavior seems to move beyond reciprocity, insofar as these people do not seek compensation, and reproductive fitness, insofar as the sacrifice of life is said to be made not only for "kin" but simply for "others."

Wilson, of course, is correct that we sometimes encounter such behavior and that we typically praise it. The question here, however, is whether his evolutionary account of our natural feelings can sustain the judgment that such actions are naturally desirable or good for those who engage in them. Wilson does posit the existence of a feeling that could give rise to such generous actions. His discussion of generosity and courage takes place in his account of sympathy, and he suggests that evolution has produced a generalized capacity for "*attachment* or *affiliative behavior*," on the basis of which human beings might engage in acts of altruism that make no contribution to their personal reproductive fitness.[60] Even if we accept the notion that natural selection alone could produce such impulses, however, it is very doubtful that they would be of sufficient strength in most people to make such self-sacrifice appear desirable. After all, presumably those human beings whose impulse toward affiliation was too indiscriminate, and therefore was not limited by reciprocity and kinship, would not have been very successful in evolutionary history. Wilson himself seems to admit as much when he notes that "sympathy" most often "does not" prompt us "to act altruistically" but merely "restrains us from acting cruelly."[61] In the end, then, Wilson's account suggests that those who practice disinterested self-sacrifice will possess a sense of affiliation of such rare strength that their actions could not be viewed as normative for ordinary human beings.

Roger Masters argues that "evolutionary principles" can "provide a foundation for the classical notion of virtue," and specifically that they can sustain a notion of "[n]atural justice" that requires of human beings "a willingness to balance one's own immediate selfish needs not only by cooperation with others in the hopes of reciprocity, but even by acts of self-sacrifice that contribute to the collective good without reciprocity." Such a standard of conduct clearly rises above mere decency, and indeed seems reminiscent of Tocqueville's concern with action in accord with those "disinterested sparks" that do credit to human nature. One wonders, however, how such notions of obligation could arise from the Darwinian account, which presents human morality as emerging from a concern with kin and from mutually beneficial reciprocal assistance. Masters suggests that such virtue is natural because large-scale society requires it. While the "principle of reciprocity can suffice as the basis of a face-to-face group," it "will not maintain the centralized state," which of necessity can only work if human beings are willing to serve the interests of other human beings to whom they are not related and with whom they may never come in contact.[62]

Such a justification of selfless virtue as natural, however, departs from the understanding of nature by which Masters's evolutionary account elsewhere suggests his normative science is to be guided. Earlier in his argument Masters indicates that "contemporary evolutionary biology" enables us "to express preferences based on a knowledge of the human behavioral repertoire."[63] On this view, it would seem, natural justice would be that which conforms to our natural proclivities. Life in large-scale societies ruled by centralized states, however, clearly does not conform to human nature as Darwinism understands it. As we will discuss in chapter 4 at greater length, the evolutionary account holds that our moral nature developed in the context of life in small, face-to-face groups; and it is in light of this understanding that Masters himself contends that the state cannot be deduced directly from human nature but only comes into existence as a result of relatively rare environmental circumstances.[64] Thus Masters's own analysis elsewhere seems to admit that life in large communities is artificial and in tension with human nature.

Because most of our evolutionary history was passed in small groups, he contends, the "extension" of human sociable interaction was "made possible only through cultural means." This development, he adds, "has put enormous strains on our species" by sharpening the natural human ambivalence towards cooperation with others.[65]

It is hard to see how the virtue required for the functioning of large-scale societies can be natural when such societies are themselves not natural. Indeed, Masters appears to admit as much when he holds that his "new naturalism" can "best be interpreted in terms of self-imposed" moral "obligations" that "depend on circumstances."[66] Such self-imposed selflessness may be necessary for society, but it is not natural in the sense of being in conformity with our evolved dispositions, which in fact could as easily be satisfied by encouraging such virtue in others while avoiding it oneself.

Robert McShea similarly betrays a desire to affirm a morality that transcends mere decency. He contends that his naturalistic account of ethics "can contain the most elevated moral insights of any of the human cultures and without loss."[67] Elsewhere, he claims that on his moral theory, which derives values from natural human feelings and which understands the good as the satisfaction of those feelings, a bad life "is literally worse than death," because it involves the defeat of "our strongest and most enduring feelings."[68] He later presents Hector's heroism, as described in the *Iliad*, as comprehensible on his moral theory. "Hector," McShea argues, displayed genuine courage "when he realized what was truly dangerous to him as a complete human being, when he fought knowing that there could be no life worth desiring, for himself or for his family or his city, if Achilles won . . . Courage is indeed the knowledge of what is truly dangerous, and that knowledge is not of an abstract concept but of ourselves, and of our total human feeling pattern."[69]

We might note, to begin with, that even if McShea's moral theory can provide a basis for it, such behavior, though admittedly impressive, is not clearly the most elevated of which human beings are capable. Hector, as McShea presents him, appears to fight on the basis of a feeling of solidarity with his community, a knowledge that his own happiness is bound up with his membership in the community under attack. Thus he seems to display what Aristotle termed citizen's courage, and not the true courage that is animated only by love of the nobility of the action itself. Moreover, it is perhaps worth noting in this context that the examples Aristotle gives of mere citizen's courage in the *Nicomachean Ethics* are from the *Iliad*. Arguably, the "most elevated moral insights" of Greek culture are to be found in Platonic and Aristotelian philosophy rather than in Homeric poetry, and it is far from clear that McShea's understanding of morality can sustain the teaching of the former.

Be this as it may, McShea's argument confronts a more fundamental difficulty. For whether or not Hector's behavior rises to the level of true Aristotelian courage, McShea's account cannot justify even such bravery as Hector indisputably displays. On McShea's understanding of the basis of ethics, our natural desires—including our sociable and moral desires—are the only ground of value, and a good life is one in which those desires are most completely satisfied. McShea contends, therefore, that human beings should make all serious decisions on the basis of deliberation about what course of action will most likely bring them the most satisfaction over the course of an entire life. Such an understanding, however, seems incapable of sustaining Hector's choice, or indeed of sustaining any decision to confront almost certain death, even in defense of the most cherished goods. After all, if the good is understood as the satisfaction of desires, then presumably premature death is to be avoided in all circumstances, insofar as it ends all possibility of satisfaction. No doubt McShea could respond that in many circumstances the avoidance of premature death will necessarily involve the frustration of many powerful and enduring natural desires, that often one can only buy one's life at the expense of one's natural desire for friends, family, and the esteem of one's community. This is, of course, true; but it does not suffice to make Hector's heroism justifiable on McShea's theory. After all, surely even some minimal satisfaction of one's natural feelings, even when accompanied by serious feeling-frustrations, is to be preferred to the end of all satisfactions (and frustrations) that death must, on McShea's theory, be understood to entail. As C.S. Lewis points out in criticizing an ethical theory similar to McShea's, even the successful traitor can eat, sleep, and copulate. Again, McShea might object that such desires as the successful coward can enjoy will be sullied by the frustrations he will necessarily impose upon himself. While this is true, at the most it can only establish that the coward's choice, which leads to some frustration and some satisfaction, is no less reasonable than the hero's, which leads through death to the absence of both frustration and satisfaction. Thus McShea's account appears at best to justify a Quisling's actions equally with a Hector's, and seems far from containing even such elevation of moral insight as one finds in Homer's presentation of public spirited manliness.[70]

While Larry Arnhart appears less interested than the other Darwinian political theorists in justifying a kind of moral virtue that goes beyond the decency that arises from our evolved nature, he does try to offer an evolutionary basis for another kind of extraordinary virtue, namely, the intellectual virtue that manifests itself in the philosophic pursuit of knowledge for its own sake. Following Darwin, Arnhart contends that man's extraordinary intellectual capacities could have evolved in the service of survival and reproduction, but that once in existence those capacities could be used to formulate "abstract ideas" with a view to satisfying the "natural desire to understand human existence."[71] The human effort to employ these capacities to satisfy this desire would then lead in some cases to the formulation of religious myths purporting to explain the cosmos and in others to the pursuit of scientific knowledge of nature. Seeking ever to be both the Aristotelian as well as the Darwinian, Arnhart concludes his treatment of these issues by suggesting that "perhaps the greatest human good, which would satisfy the deepest human desire, would be to understand human nature within the natural order of the whole," thus seeming to endorse Aristotle's presentation of the life of philosophy as the best for man.[72]

Arnhart's position is subject to essentially the same objections as Wilson's. Because knowledge of the natural world would be useful with a view to reproductive fitness, it is perhaps not unreasonable to suppose that evolution could generate both our seemingly unmatched intellectual capacities and a curiosity about the world around us. For philosophy to be the best way of life for human beings, however, it would be necessary—at least on a theory like Arnhart's that determines what is best for human beings on the basis of their evolved desires—for the disinterested love of knowledge to be, as Arnhart indicates, the strongest human desire. It is extremely doubtful that this would be the case for most people. After all, presumably those human beings whose intellectual desire was insufficiently directed toward knowledge useful with a view to reproductive fitness would not have been very successful in evolutionary history. We would expect that most people, then, and even those of superior intelligence, would seek practical knowledge with a view to power or status and that only a few would seek wisdom about the whole for its own sake. Thus the philosopher, no less than the heroic altruist, would be a kind of natural curiosity whose understanding of the good would not be normative for most human beings.

Francis Fukuyama acknowledges the existence of "virtues like courage, daring, statesmanship, and political creativity," ethical excellences that both show themselves in and are necessary for the creation of large-scale political communities. He remarks that such virtues are not only (empirically) "less frequently observed" but also (normatively) "greater" than those that emerge spontaneously from our natural sociability.[73] Fukuyama seems to manifest a desire to affirm the nobility of spirit displayed in such difficult virtues, but like the other Darwinians presents these more elevated virtues as distinct from those that are rooted in human nature. If they do not arise from our nature, however, it is difficult to see on what grounds one could reasonably view the possession of such virtues as preferable to their absence.

Fukuyama might contend, like Masters, that such virtues derive their justification from their necessity for the success of large-scale political order; but again, since large-scale political communities are themselves not rooted in our nature, it is unclear why they, or the virtues on which they depend, should be naturally desirable. In response, Fukuyama might point out that many good things depend on large-scale order. On the basis of the spontaneous order that arises from our natural sociability, Fukuyama indicates, there will be "no division of labor, no impersonal market, no economies of scale, no legal guarantees of property rights and therefore no long-term investment," hence no "high art" or "scientific research."[74] Such an argument, however, indicates that the noble virtues are useful rather than inherently choiceworthy, and from the standpoint of any given citizen such an understanding points not to the importance of cultivating excellence but to the need to encourage it in others so that one may enjoy its fruits.

CONCLUSION

The Darwinian political theorists offer, as we have seen, a reasonable argument that our moral and sociable desires are rooted in human biology. Nevertheless, their account of human nature tends to reduce morality to mere decency, and even when they try to justify a virtue that rises above this level they fail to do so convincingly. Thus their scientific justification of morality fails to provide a basis for overcoming democracy's decent materialism, and religion retains, at least from the standpoint of Tocqueville's concerns, its public importance. In the next chapter, we turn to the question whether Darwinism can sustain universal morality, and hence whether religion is necessary, as Tocqueville contends, to answer the threat to freedom posed by the tyranny of the majority.

4

Moral Universalism

I F DARWINISM FAILS TO OFFER a lofty morality that could en-
noble modern society by restraining democratic materialism,
can it at least sustain a universal morality that could correct
the democratic inclination toward tyranny of the majority? At first
sight, the Darwinian account appears equal to this second Tocquevillian task. After all, as we have observed, it establishes the natural
foundations of human feelings promoting sympathy and reciprocal
justice.

This alone does not suffice for Tocqueville's purposes, however.
The democratic isolation or individualism he deplores is not an utter
egotism, but a restricting of one's sympathies to a narrow circle of
family and friends. To correct the danger Tocqueville sees, Darwinism
would need to show a natural basis for expanding the sense of moral
obligation beyond such limits. It would have to teach, as religion
teaches, that there are some obligations owed to all human beings as
such. Here it fails: the evolutionary account of the origins of morality prevents the Darwinian moral teaching both from reaching very
high and also from extending very far. In the end, the Darwinian

account of human nature makes groundless not only the aspiration for a decent international order, but also any aspiration for a decent domestic order in a large society.

According to the Darwinian account, human morality evolved in the context of cooperation within small groups. "Throughout most of human evolutionary history," Arnhart observes, "human beings lived in hunter-gatherer groups" in which "those who acted for selfish interests contrary to the interests of the group were punished." In such an environment, natural selection would favor the emergence of "cooperative dispositions" in individuals to promote the well-being of the group.[1]

As the Darwinian political theorists generally recognize, this theory implies that moral universalism is not rooted in human nature. Fukuyama, for example, expressly states that neither our biological nature nor the "spontaneous order" that humans produce through "decentralized bargaining" can give rise to "moral universalism," or "moral rules that apply to human beings qua human beings."[2] Similarly, Arnhart notes that for Darwin the prevalence of war among tribal peoples is no refutation of natural human sociability, because "the social instincts never extend to all the individuals of the same species."[3] James Q. Wilson holds that "the human proclivities and social experiences that give rise to our moral sense tend to make that sense operate in small groups more than in large ones, to say nothing of embracing mankind as a whole." Therefore, "we are by nature locals, not cosmopolitans."[4]

Indeed, the Darwinian account holds that morality emerged to promote success in the conflicts between groups that prevailed during the period that our nature evolved. Arnhart contends that "natural selection favored those moral norms that allowed members of a group to cooperate with one another" in order to "compete with other groups."[5] Universal morality appears to be not only not rooted in human nature but actually contrary to it. Natural human morality (that is, in-group morality) is itself the basis for the most egregiously immoral behaviors between groups.

Spontaneous or natural order, Fukuyama indicates, gives rise to closed communities; and the natural morality or proclivity for cooperation that sustains such communities is often the basis for "shocking immorality" between communities.[6] After all, as he points out, "the Germans who carried out the Holocaust" could not have done it without a considerable amount of "loyalty" to, and willingness to sacrifice for, their own community.[7] Because "biological dispositions favoring social cooperation" result not in "moral universalism" but in "the selfishness of small groups," adherence to moral obligations owed to "humanity as a whole" requires that we to some extent dispense with natural morality, that we "violate deeply felt norms of loyalty and reciprocity to our particular groups."[8] Fukuyama is, as we will see, emphatically committed to a universal notion of human dignity. According to his Darwinian understanding, however, respect for the dignity of others exists *by nature* only within small groups, which is to say that *human* dignity, the acknowledgement of the dignity of humans as such, does not exist at all by nature.

The disturbing extent of nature's indifference even to behavior most flagrantly opposed to universal human dignity is suggested by Fukuyama's at first apparently harmless comment that "states can't make individuals follow norms that violate important natural interests or instincts."[9] Consider this statement in light of Fukuyama's reference, only one page earlier, to the various murderous totalitarianisms of the twentieth century. Apparently, on this account, the "systematic genocide" that the Soviet and Nazi systems required their citizens to perform, and that many of those citizens more or less willingly carried out, violated no "important natural interests or instincts."

Other Darwinian theorists acknowledge similar implications. Wilson notes that our sense of sympathy for outsiders is weak "precisely because it arises from the same sociability that makes us want to go along, be liked, follow orders, win approval." Consequently, "[o]ur capacity for being decent persons and one of the sources of our deepest depravity are, at root, one and the same."[10] Robert McShea points out that biology's refutation of the belief in "universal egotism" does not necessarily suggest a "more optimistic view of human nature." For when we consider "dedication to humane causes" we have "seen only half of 'altruism,'" while the "other half" involves such things as patriotic imperialism and a nepotism that "makes a mockery of many systems of justice. Even racism can be seen as a kind of group solidarity."[11] Roger Masters points out that war and conquest are not "hard to explain from the perspective of inclusive fitness theory," since defeating and enslaving another people has "enormous benefits for the kin of all individuals among the victors."[12] Evolutionary biology thus suggests that the mass murder of those outside one's group is closer to being the law of nature than a violation of it.

Darwinism, then, cannot provide a basis for moral universalism and therefore cannot sustain what we may well take to be some of our most self-evident moral judgments—for example, that Nazi foreign policy, which aimed to exploit or exterminate inferior races, was unjust. We might, moreover, reasonably question an account of morality that fails to sustain some of our most deeply and widely held moral convictions. One might defend the Darwinian account, however, by contending that this supposed defect is only a failing from the standpoint of a teaching, like Tocqueville's, influenced by the moral universalism of Christianity. That is, to criticize Darwinism for its failure to support universal morality presupposes that universal morality is important, that universalism is an essential element of a complete morality. But perhaps it is not. Indeed, given what evolutionary biology claims to reveal about the origins and nature of our moral inclinations, the Darwinian might dismiss as utopian the call for moral universalism.

For their own part, however, the Darwinian political theorists deny themselves this defense by expressly embracing universal morality. They recoil from the consequences of their teaching and try to find some way to support moral universalism, despite its apparent lack of a grounding in nature as they understand it. These attempts fail, and thus their own arguments bear implicit witness to their theory's inability to sustain their own moral aspirations.

UNIVERSALISM AND HISTORICISM

Fukuyama evidently wants to affirm the existence and propriety of universal morality. He notes with disapproval the irrationality of the traditional bases of community, such as ethnicity and religion, and their proclivity to generate conflict. In contrast, he praises the Enlightenment for its attempt to erect "a political order based on the universal recognition of human dignity" which alone can "avoid these irrationalities and lead to a peaceful domestic and international order."[13] As we have seen, Fukuyama also recognizes that our evolved moral nature is not supportive of but in fact hostile to universal morality. He seems nonetheless oddly complacent about the continued persistence of universal morality. Even as he notes that western societies have suffered a decline in the ordinary morality of daily life, he expresses confidence that those same societies will be able to continue "to enforce universalistic principles of individual rights and citizenship."[14] What force, apparently greater than nature itself, can sustain this confidence?

The immediate answer seems to be, "the modern state," since Fukuyama notes that our universalistic morality is enforced by "plenty of hierarchical authority."[15] Ultimately, this answer is insufficient, as the hierarchical authority of the modern state is no more rooted in our evolved human nature than is universal morality itself. In the end—and this perhaps comes as no surprise, in light of Fukuyama's earlier work, especially *The End of History and the Last Man*—*The Great Disruption* hints that history or historical progress provides the ground for universal morality.

Fukuyama credits revealed religion with the introduction of universal morality. Religion first pushed the sense of moral obligation beyond the small group, and we therefore "owe to religion the fact that it is civilizations rather than families or tribes that are today the basic unit of account." Moreover, "it is religion alone that first" proposed "mankind itself" as the "final community within which its moral rules should apply."[16] Fukuyama further contends, however, that universal morality is now "an inextricable part of the moral universe created by religion."[17] Although he presents revealed religion as irreversibly on the wane, he suggests that universal morality has now been embraced by secular philosophies such as modern liberalism, and that "the cultural patterns it established long ago continue to play decisive roles in shaping contemporary trust relationships."[18] Today, it seems, liberal politics bears "nearly the entire weight" of supporting "the principle of the universality of human dignity." At this "it has done a remarkably good job" and, moreover, will almost certainly continue to do so. Fukuyama contends that we can "expect a long-term progressive evolution of human political institutions in the direction of liberal democracy."[19]

It is not clear that this solution to the problem of universal morality is compatible with Fukuyama's commitment to the Darwinian account of human nature. If our moral nature is not oriented toward universalism and is in fact hostile to it, why should we expect history to follow such a long-term trend? Fukuyama might respond that such a trend is to be expected because a universal liberal political order best satisfies the natural human desire for comfort and security. In fact, he suggests this in *Our Post-Human Future*. Such a solution carries serious moral and theoretical costs, however. As Fukuyama points out in *The Great Disruption*, "liberal societies buy political order at the expense of moral consensus," or to put it another way, they purchase peace at the price of moral seriousness. Thus they provide no "moral guidelines" beyond the "universal obligations of tolerance and mutual respect." They hold out no elevated conception of virtue toward which we must strive, but insist only that citizens refrain from violating each other's basic rights.[20] To seek to establish universal morality on such low foundations is to surrender to the decent materialism that Tocqueville thinks democracy must transcend and, furthermore, to repudiate one of the central concerns of the Aristotelianism whose credibility Fukuyama's science of politics claims to restore.

In any case, the historical evidence does not obviously support Fukuyama's optimism about the future of universal morality. For while such a trend in the direction of a liberal respect for human rights is clearly discernible, its permanence is not yet certain. The modern state has been turned against universal morality many times in recent memory, and it is not clear what, besides a relatively peaceful trend in the last fifty years, should give us confidence that this will not happen again. Finally, even if Fukuyama is correct that History has foreordained long-term progress in the direction of respect for universal morality, this trend may be taken as an empirical fact with no necessary moral implications. That is, while the triumph of human rights may be inevitable in the end (and Fukuyama would doubtless admit that that end may be a long time coming), that is no reason why those with power and opportunity should decline to profit from tyranny while they still can.[21]

COMMON NATURE AND CONFLICTING DESIRES

Robert McShea likewise endorses the expansion of the sense of moral obligation beyond the narrow scope within which it initially evolved.[22]

He calls for the "production of more inclusive and generously imagi-native conventions" and implicitly praises the sense of "solidarity with the rest of the human race."[23] Moreover, he indicates that these "inclusive conventions" can be grounded in "human nature," properly understood.[24] Given our common human nature, McShea argues, moral consensus—and hence, presumably, justice among humans as such—is in principle possible. "Because the elemental feelings and their relative strength are similar in all or most of us . . . there can be only contingent, accidental, reasons for failure to come to value and action agreement." Such contingent obstacles include "ignorance, prejudice, misunderstanding, present passion, and perceptual and imaginative inadequacies." The existence of such obstacles "represents merely a practical difficulty, however overwhelming, not a principled objection."[25] McShea accordingly presents his evolutionary human na-ture theory as a basis of human solidarity. "Human nature theorists," he contends, should, because of their appreciation of our common humanity, be "less prone to racism or ethnocentrism, or to any of the other ways of" pitting some human beings against others.[26]

As was the case with Fukuyama, however, McShea's own Darwin-ian principles undermine the universalism he desires to support. The thrust of his argument is that by affirming our common nature, the evolutionary account encourages fellow feeling among humans be-ings as such. Yet such feelings do not necessarily follow from a mere recognition of our common humanity. As McShea points out many times, on his evolutionary theory, emotions or feelings, not cognitive knowledge, are the source of moral values. For example, he contends that in the case of abortion what matters is not the abstract question of the humanity of the fetus but instead "how we feel about it and how that feeling fares in its conflict with other feelings."[27] Surely the same reasoning would apply with reference to the possibility of universal morality: what matters about those outside our social group is not our common humanity, but how we feel about them.

How are we likely to feel about them, on Darwinian principles? McShea's argument points to the uncomfortable answer: the "most deadly struggle in nature is not between species . . . but among conspecifics."[28] Presumably, in a sociable species like humans that struggle will take the form of war among groups. McShea later makes this explicit, observing that our present, natural "psychological profile" or "feeling pattern" evolved in a context in which "[o]ther groups were feared and hated, and those feelings worked to strengthen the internal cohesion of the group."[29]

McShea contends that in light of our common human nature there are only accidental reasons to disagree. Yet this observation loses its intended practical or moral force when one realizes that, according to McShea's own account, our very nature magnifies the psychological or emotional importance of those "accidental" differences to the point that they tend to eclipse our common humanity. McShea lists "prejudice" among the accidental causes of division, but then must admit that we are inclined by our very nature to be prejudiced. We prefer the well-being of the members of our group, even at the expense of the well-being of those outside it, despite the fact that it is somehow accidental or contingent that some of use belong to one group and others to another. The circumstance that this child is mine and that one is yours and that resources are scarce is an "accident" in comparison to the universal phenomenon of parental care, but such accidents are, not despite but because of our common nature, the source of much conflict. In sum, a common nature need not give rise to common aims, and shared feelings are useless as a source of cooperation when the objects of those feelings are of necessity different. Achieving a greater realism and consistency than McShea, Richard Alexander notes that modern biology shows that conflicts of interest among humans are not accidental but stem from the fact that each person, having an individualized set of genes, has an individualized set of interests.[30]

WILSONIAN UNIVERSALISM

Like his Darwinian colleagues, James Q. Wilson is drawn to the idea of moral universalism. By characterizing intergroup oppression as a manifestation of our "deepest depravity" he implies the existence of moral standards among peoples.[31] Nevertheless, he acknowledges that on his own Darwinian theory of human nature such universalism seems to have no place and therefore is something of a puzzle.[32] Wilson attempts to solve that puzzle, contending that moral universalism arose in Europe through a fortunate set of historical developments conditioned in part by the moral teaching of the Catholic Church. Of significance here is not so much the general universalism of Christian ethics as the understanding of marriage as a sacrament. This notion, Wilson contends, required that marriages be consensual, which freed them from clan control, which in turn created relatively autonomous families, thus eroding the traditional tribal mentality and giving rise to a new spirit of individualism. And "[i]ndividualism implies universalism, since if each person is morally equivalent then all peoples are morally equivalent."[33]

Wilson also suggests that moral universalism may have some basis in our evolved nature. Natural selection, he contends, could have generated in human beings not a "simple desire for reproductive success" but instead "a generalized trait that both encourages reproductive fitness and stimulates sympathetic behavior." Such a broad capacity for "attachment or affiliative behavior" would then be the basis for a natural human concern with the interests of non-kin.[34] While this impulse to attachment would have been constrained by the tribal circumstances in which humans lived throughout most of their evolutionary history, the development of larger-scale communities has allowed it to show its full power. Wilson remarks in this context that the "strength" of the moral senses comes to light "when circumstances permit their fuller expression and wider effect."[35]

This attempt to ground moral universalism in human nature is problematic in light of the evolutionary principles Wilson embraces. First, it is not clear why natural selection would generate a generalized capacity for affiliation in order to promote reproductive success when narrower sympathies would surely suffice. Indeed, when one recalls that the evolutionary environment was characterized by constant conflict among small bands, such a general capacity for affiliation with human beings as such, with people outside the group, would appear to be positively detrimental and doomed to extinction.

It is in this light that we can appreciate the strangeness of Wilson's remark that the full strength and scope of the moral sense is manifested only when circumstances permit it. Here Wilson evidently refers to the emergence of a large-scale public order as the condition for the full flowering of the moral sense. Thus his argument suggests that our natural moral sense comes completely to life only when placed within an environmental context unprecedented in human history. According to Darwinism, however, human nature, including the moral sense, is wholly the product of adaptation to circumstances, and specifically to circumstances radically different from those in which widespread sympathy is likely. Wilson's remark thus suggests that there is some element of our nature somehow not shaped by that early environment, which on evolutionary principles is impossible.

Whether or not one could devise a plausible evolutionary argument for the natural status of a generalized proclivity for attachment, Wilson's own argument suggests that such an inclination, even if it exists, is too weak to sustain a universal morality. He comments that the "moral sense for most people remains particularistic" while only "for some, it aspires to be universal."[36] Indeed, Wilson seems to give up on the notion of universal morality when he notes that affiliation, however far one may seek to expand its scope, ultimately requires social "boundaries," a "we" defined in contrast to a "they."[37]

EVOLUTIONARY EQUALITY

Roger Masters also approves of universal morality and seeks to find some basis for it in his Darwinian theory of human nature. He expresses his disapproval of genocide and forced expulsions, and he criticizes modern political philosophy for its emphasis on "abstract claims"—such as the "rights to life, liberty, and estate"—"that produce social conflict without providing natural grounds for self-restraint in the exploitation of either the physical environment or other peoples." While the "natural rights doctrines of Hobbes and Locke gave rise to colonialism" and the gross exploitation of "the indigenous population of North America," evolutionary naturalism, Masters contends, supports a view of "natural justice" that can "recognize the claims of others."[38]

What, then, is this natural justice, and how is it derived from Masters's Darwinian theory? "The first principle of natural justice that should attract our attention," Masters argues, "is the respect for other human beings entailed by our common humanity." If, as evolutionary science suggests, "the phenotype," or the organism itself, "is merely the vehicle by which genes replicate themselves, every human being is equal in a more profound sense than has been realized in most secular political teachings." After all, "who can know which of us carries a mutant gene that is a valuable adaptation to a future environment and will someday spread throughout the human gene pool?" No human genotype "can ever be said, a priori, to be better than another, if only because the natural selection that will test differential survival lies in the future."[39]

In the first place, one might note that the observation on which Masters hopes to base this obligation to respect individuals for their mere humanity could be restated, with as much accuracy, as follows: "anyone we encounter might be, but probably is not, carrying a gene that will one day prove necessary to the survival of the species." Alternatively, one could say with equal truth that any person one meets might be carrying a gene that will one day prove initially useful but ultimately disastrous for the species. When phrased in these less optimistic, but no less truthful, ways, Masters's observation seems to lose considerable force as a justification for respect for individuals.

More important, here Masters's argument conveniently ignores other implications of his Darwinian science, namely, that we are evolved to value some genotypes over others—especially those to whom we are genetically related or with whom we are closely associated—not to care particularly about the long-term prospects of the species as a whole. Earlier Masters had suggested that his normative science of human nature would express preferences or value judgments based on "knowledge of the human behavioral repertoire" as it has "evolved over millennia."[40] When justifying respect for individuals, however, Masters appeals not to our evolved behavioral propensities—or, as the other Darwinian political theorists might put it, our natural feelings—but to the well-being of an abstraction: the future human species. But the feelings to which evolution gives rise give scant support to a concern for the well-being of the species as a whole, much less an interest in the species extended into the distant future. On the contrary, those feelings might in some cases provide a powerful incentive to disregard the humanity we share with others. After all, the North American colonists that oppressed the Indians were acting on perfectly natural desires. As Masters himself notes, conquest and slavery are easy to explain on the basis of inclusive fitness, since the spoils will enrich the kin of the victors and masters.[41]

MORAL TRAGEDIES

With reference to universal morality, Arnhart seems to be generally more consistent and realistic than his Darwinian colleagues.

More than the others, he tries seriously to take human nature, as revealed by Darwinian biology, as his sole standard, resisting the importation of moral concerns that cannot be explained in light of our evolved desires. He criticizes Darwin himself for his moral "appeal to universal humanitarianism" which, Arnhart argues, "can only be explained as a utopian yearning for an ideal moral realm that transcends nature" and which therefore "contradicts" Darwin's scientific "claim that human beings are fully contained within," and hence can be fully explained in terms of, "the natural order."[42] On the basis of the Darwinian account of the origins of human sociality and morality, Arnhart contends, it is far more realistic to believe that "war" is a permanent feature of the human condition because "it is rooted in the natural desires of human beings as political animals."[43] Indeed, he lists war among the twenty natural desires, remarking that "[h]uman beings generally desire war when it seems to advance their group in conflicts with other groups."[44]

Ultimately, Arnhart suggests that the limits of our natural morality can in some cases render irrelevant the question of the justice of war. If "human beings are not bound together by a universal sentiment of disinterested humanitarianism," Arnhart argues, "then deep conflicts of interest between individuals or between groups can create moral tragedies in which there is no universal moral principle or sentiment to resolve the conflict." When confronted with conflict, "we must either find some common ground of shared interests, or we must allow for an appeal to force or fraud to settle the dispute." The only other alternative, which Arnhart expressly rejects as not "realistic," is the invocation of "some transcendental norm of impartial justice (such as Christian charity) that is beyond the order of nature."[45]

Ultimately, however, Arnhart does not fully embrace the consequences of this realism and thus involves himself, like the other Darwinian political theorists, in the difficulty of maintaining moral positions that his scientific principles cannot sustain. We will return to this point later in this chapter. For the present, suffice it to observe that even here, in offering a uniquely candid statement of the "realism" required by evolutionary morality, Arnhart himself seemingly cannot avoid implicitly indicating a concern with universal morality that his Darwinian account cannot explain. The very desire to characterize such situations as "tragic"—as somehow bad or un-fortunate—implies a moral perspective that transcends our evolved feelings. If our moral feelings can be explained wholly as the product of nature, would we not regard such dilemmas not as "tragic" but simply as the way of the world, indeed, not even as dilemmas at all, insofar as they can be resolved (possibly in our own group's favor) through force or fraud?

UNIVERSAL MORALITY AND TYRANNY OF THE MAJORITY

Tocqueville, as we have seen, is not concerned primarily with inter-national justice, but with the maintenance of decent and just condi-tions within modern democracies. His concern with the narrowness of the morality fostered by democracy arises from his concern that people within modern democracies should treat each other justly. Darwinian morality, then, is politically problematic not only because of its indifference to international justice, but also because of its inability to support even a decent domestic order. The large scale of modern societies far exceeds that of the communities in which human nature evolved. There appears, then, to be no natural reason to treat justly those fellow members of one's political community who are outside one's own circle of kin and friends. That is, there is no reason why we should not desire domestic politics to be about inter-group exploitation, so long, at least, as the prospects are good that our own group will be the beneficiary. To use Tocqueville's ter-minology, there is no principled reason for majorities not to engage in tyrannical behavior.

The Darwinian political theorists themselves acknowledge this difficulty. Arnhart, for example, points out that the emergence of large-scale societies "created opportunities for cheaters," those who accept the benefits of society without helping to shoulder its burdens, and that this in turn provided the incentive for the establishment of "sovereign rulers capable of enforcing obedience from all."[46] In other words, because of the conditions under which morality evolved, there is no reliable natural impulse toward cooperation in the large group, and this deficiency must be supplied by fear of the coercive power of the state. Similarly, Masters contends that the modern, bureaucratic state cannot be deduced from our natural sociality. He points out that something like a Freudian ambivalence regarding cooperation or altruism is a predictable consequence of life in large societies, in which human beings are alienated from the natural conditions in which cooperation would be experienced as much less problematic.[47]

Not content to leave it at this, however, the Darwinian political theorists also try to find a way to sustain domestic justice on the basis of their evolutionary account of human nature. They condemn manifestations of what Tocqueville would call majority tyranny as contrary to natural right and recommend as naturally just systems designed to protect the interests of the weak. Once again, however, such attempts fail and thus suggest that Darwinian morality is problematic even from the standpoint of those promoting it.

DARWINISM AND CONSTITUTIONALISM

According to Roger Masters, Darwinian science points to a conception of natural justice that suggests the desirability of constitutional democracy. Just as Aristotle's science of politics "combines clinical if not scientific objectivity with the capacity to make discriminations based on the natural capacities of our species," so, Masters contends, "contemporary evolutionary biology" can "express preferences based on a knowledge of the human behavioral repertoire." Some "social systems," he continues, "are closer" than others "to the central tendency of human behavior produced by our evolutionary past," and therefore more in accord with natural justice.[48]

Once again, human nature evolved under conditions in which human beings lived in small hunter-gatherer bands of perhaps fifty to one hundred members. While such groups certainly were governed by dominance hierarchies, they also were sufficiently egalitarian that there was widespread opportunity to engage in the behavior that Masters terms "threat" or "voice," on the one hand, and "flight" or "exit," on the other.[49] That is, most members could assert their interests with some reasonable chance of modifying the group's behavior and, if dissatisfied with their treatment, leave the group. Masters therefore concludes that "constitutional regimes combining the rule of law and widespread political participation come closer to fulfilling the natural potentials of our social nature than do autocratic or tyrannical governments."[50] The "balance of voice and voluntary exit characteristic of hominid bands prior to the development of large-scale agriculture is more closely approximated in the constitutional democracies of the West than in autocratic or totalitarian regimes." While the former maintain "avenues for widespread social and political participation" and limit forced exit, or coercive removal from society, "to specified legal procedures of a public nature," the latter, by excluding all but a few from "effective political participation" and minimizing "voluntary exit," tend to "prevent large sectors of the population from engaging in behaviors typical of the human social repertoire."[51] Masters concludes that the "democratic political processes associated with republican or constitutional forms of government, like the informal decisions in hunter-gatherer bands and other face-to-face groups, are 'naturally right' or healthy for human societies."[52]

Masters's case comes down to this: constitutional democracy is better—in the sense of more emotionally satisfying, because more responsive to our natural psychological desires—for a population of human beings than is autocracy or totalitarianism. This observation is perfectly true but nonetheless morally irrelevant, at least on a Darwinian account that seeks to establish moral obligations only on our natural feelings or, as Masters puts it, evolved behavioral repertoire. By Masters's own admission, an impulse to care about the well-being of humans generally is not part of our natural repertoire of behaviors. He asserts that the capacity to identify with a society beyond the small group is "clearly cultural."[53] On the other hand, there are powerful natural feelings that lead us in the direction of denying to others the conditions of their well-being. Again, Masters himself notes that conquest and slavery are perfectly explainable in terms of inclusive fitness, since the benefits of such injustice can be given to one's kin and intimate associates.[54] Elsewhere, he points out that "repressive or totalitarian regimes" are often caused in part by "attempts to prevent downward social mobility by members of social classes whose socioeconomic status is insecure."[55] Once again, given the reproductive fitness value of high socioeconomic status, this is, on Darwinian grounds, perfectly reasonable behavior. In sum, though it may be true that constitutional democracy is good, and tyranny bad, for the population at large, there is no natural reason for the tyrannical faction to care that its flourishing depends on denying the conditions of well-being to other groups.

Robert McShea tries to defend the same political position as Masters and falls into essentially the same difficulties. McShea contends that "[o]ur knowledge of our nature suggests that at any level of technology or culture, under any possible regime, and without regard to the size of the society, there will be leaders and led, elites and masses."[56] The relevant question to be addressed in framing a decent society, then, is not how to abolish hierarchy or establish egalitarian conditions, which is impossible, but instead how to restrain elites so that non-elites will be reasonably safe from injustice. McShea recommends combining a public morality sufficiently internalized in elites that they will to some extent check themselves, with a political system that compels elites to compete for public authority so that they will to some extent check each other. Such arrangements, McShea argues, seem "to afford the best protection for non-elites. Elites are mildly inhibited by internal restraints, checked a little by public opinion, and so divided against themselves that some elite element will always find it to its interest to defend the social contract and the interests of non-elites." "The sensible strategy," then, "for those who favor the cause of non-elites would be to promote the pressure of constitutional restraints upon as a large a number of independent and competing elites as possible and to support the cultural solidarity that is the only support of the constitution."[57]

McShea's observations here are perfectly sensible, but they beg an important question: why should we favor the cause of non-elites as opposed to the cause of the elites? There is, of course, an evident self-interested answer for most of us. Not being members of the elite, we favor its restraint out of a natural concern for our own well-being. The more difficult question, however, is this: why should elites favor the cause of non-elites, or why should some particular elite support such a system if it thinks can succeed in exploiting the non-elites and suppressing possible competitor elites? This question must be answered if McShea's proposed system is to be regarded as in any objective sense better than the various possible alternatives that do not make a priority of protecting the interests of non-elites.

According to McShea's theory, the answer to this question, and indeed the answer to any question of value, can only come by way of some natural feeling that obligates the person experiencing it. Is there some natural sentiment that interests members of elites in the well-being of those of lower status who are not their own kin and whom they do not know personally? McShea indicates the existence of such a feeling when he contends that it is necessary for an elite-restraining constitution to be "supported by a secular civic morality" which in turn "assumes and exploits innate feelings of human solidarity."[58] Yet the Darwinian account of the origins of human sociability and morality casts grave doubt on the existence of such feelings, or, if they do exist, on their strength relative to other evolved feelings more directly supportive of inclusive fitness. That such a system as McShea describes is better for most people is perfectly true, but that does not make it naturally just in any sense that obliges those who can benefit from some other more exploitative arrangement.

SLAVERY AND THE MORAL SENSE

Despite his desire to adhere to a moral realism grounded solely in the evolved desires of human beings, Larry Arnhart ultimately moves in the direction of affirming universal morality, and thus moves beyond the morality that his Darwinian principles can support. He does this in his argument against human slavery, to which Arnhart devotes the longest chapter in *Darwinian Natural Right*. While Arnhart contends that slavery is contrary to human nature and therefore unjust, a careful consideration of his arguments indicates that his evolutionary account of human nature cannot sustain his right-minded repudiation of this most egregious form of majority tyranny.

Slavery, Arnhart contends, involves "the unjust contradiction of treating human beings as if they were not human."[59] To suggest that a certain action is morally wrong because it involves a contradiction —that is, an irrationality—might seem to imply that justice is to be determined with reference to abstract reason. Arnhart forthrightly denies this, however, insisting that we must appeal to human nature, and more specifically to our natural desires, as the basis of morality. In his critique of slavery, then, Arnhart means to suggest that slavery involves a contradiction of human desires.

Arnhart's argument points to two ways in which the master-slave relationship involves an emotional contradiction. First, he notes that slavery contradicts the slave's desire not to be enslaved. Once again, we have evolved to engage in and be satisfied by reciprocal cooperation, from which the one-sided exploitation of slavery radically departs, provoking a natural resistance on the part of the exploited. All normal human beings are "naturally inclined to assert their independence" against a slave system, which is "naturally bad because it frustrates a natural desire to be free from exploitation."[60]

No doubt this is true: slaves generally do not like being slaves. But why should the master care? As Arnhart admits, slavery arises precisely because it appeals to some natural desire of those exercising mastery. There is a "natural self-love" that "inclines some human beings to exploit their fellow human beings whenever they can get away with it," and one manifestation of this is the "natural desire" of "masters" to "exploit slaves."[61] In this context, Arnhart's appeal to the natural human desire to be free from exploitation appears as an (impermissible, on his account) reliance on abstract reason. That is, his argument, by appealing to what is good for human beings generally without explaining why the humans in question (masters) should care about the well-being of certain others humans (slaves), treats human nature as an abstract principle that obliges observance regardless of the feelings of those on whom it places moral demands. To appeal to the feelings of others gets us nowhere absent some reason we should care about the feelings of others.

Arnhart does offer such a reason, however. Because the desire to be free of exploitation is no less natural than the desire to exploit, slaves will resist. Slavery is therefore naturally unstable. "Like other social animals," Arnhart observes, "human beings are naturally inclined to exploitation through coercion and manipulation. But human beings are also naturally inclined to detect and punish exploitation."[62]

Thus the contradiction inherent in slavery is manifested in the fact that it always rests on force.[63] As Arnhart observes, conflict often accompanies attempts by humans to use even nonhuman animals as tools. But "[w]hen human beings try to hold other human beings as property, the struggle of wills becomes more intricate, because this form of property can defend its interests with as much emotional strength and intelligent calculation as its owners do in defending their interests." Slavery, therefore, always creates "an unremitting contest between the master's desire to exploit the slave and the slave's desire to resist such exploitation."[64]

This line of argument, however, establishes nothing more than that slavery is hard to maintain. But hard is not impossible, and given the gains to be realized, potential masters might well be inclined to hazard the difficulties. Indeed, Arnhart admits that "slavery has often been deeply rooted in the customs and laws of various societies," which seems to indicate that it is possible to some extent for masters to overcome the problems it poses.[65] To be sure, even a successful slave system will involve the tension and instability Arnhart mentions, but it is not clear why such problems should render the institution unnatural. After all, Arnhart's argument indicates that the conflict between the desire to exploit and the desire to detect and avoid exploitation exists throughout living nature.[66] It is not clear, then, why the relationship between master and slave is any more unnatural than the relationship between predator and prey. Nor is it clear why, on Arnhart's principles, this particular form of human exploitation should be unnatural when other forms are considered natural. After all, one's opponent in war also resists being conquered, yet this does not stop Arnhart from listing war as responsive to one of our natural desires and judging it a natural and permanent part of the human condition.[67]

Perhaps because of the limited moral relevance of the fact that slavery contradicts the slave's desire to be free of exploitation, Arnhart does not rely on this argument alone but points in addition to a second contradiction that slavery involves: it contradicts the feelings of the master himself. Here Arnhart is on ground more appropriate to his general approach, which, as we have seen, posits our natural feelings as the sole source of moral obligation. Slavery, Arnhart contends, "violates those moral sentiments that arise from sympathy and reciprocity." Therefore, the "emotional constitution of human beings," even those who are not themselves slaves, "inclines them to feel disgust" at the "injustice" of slavery.[68]

This is not, however, the whole story. As Arnhart has already admitted, there is a natural desire to exploit that is opposed to our feelings of reciprocity and sympathy. Why, then, should we prefer one to the other? If one enslaves another, one will experience painful feelings, the guilty pangs that attend violations of sympathy and reciprocity. On the other hand, if one refuses to enslave another when a reasonable opportunity presents itself, one will have to mortify the desire to exploit and will experience the pain of the missed opportunity for easy gain.

Is one to reject the desire to exploit on the grounds that it is naturally weaker than our sense of sympathy and reciprocity and therefore that its indulgence will result, on balance, in more psychic pain than its rejection? This hardly seems credible, especially when one reflects that slavery is not practiced for the master alone. Slavery benefits the master's kin and friends, for whom he will naturally cherish powerful feelings, while the burden of its injustice falls on mere generalized human beings, those beyond the master's group and defined by him as outsiders, for whom he will by nature have no feelings or only weak feelings. As Arnhart himself admits, a "person's humanitarian sympathy for strangers will almost always be weaker than his egoism, his nepotism, and his patriotism."[69] It seems that, at least on Darwinian grounds, slavery pays the master more in emotional satisfaction than it costs him in emotional dissatisfaction.

As a response to this problem, one might contend that, though weaker than our desire for nepotistic exploitation, our sympathy and sense of justice is sufficiently strong that when we enslave others we will get, in addition to the gratification of our desire to exploit, a constant sense of psychic pain arising from our guilt. Thus slavery is ultimately not worth the emotional trouble it brings the master: he might as well pursue some less exploitative means of gain, one that results in the flourishing of his interests but avoids the sense of disgust that arises from violating the norm of reciprocity.

Once again, the Darwinian account of human nature appears to undercut the possibility of such an argument. Arnhart notes that because slavery violates our deeply rooted moral feelings about how human beings should treat each other, slaveholders typically try to deceive themselves and others by denying the humanity of slaves.[70] He expresses confidence that such attempts must ultimately fail, and thus the injustice of slavery must continue to irritate the consciences of masters. But the evolutionary account of human nature indicates otherwise. James Q. Wilson observes that we seem to dehumanize those we treat unjustly.[71] There is good reason, moreover, to view this capacity for self-deception in the service of immoral self-interest as every bit as natural as our capacity for sympathy for those close to us. If morality, and our emotional constitution generally, emerged under conditions of intra-group cooperation in the service of ruthless inter-group competition for resources, then a certain natural capacity for hardness, for dehumanization of the outsider would seem to be adaptive. (Robert Wright deals with the natural basis this capacity in *The Moral Animal*.)[72] Again, we are left with the question: what natural reason is there that we should reject the natural proclivity for dehumanization in the service of unjust gain in favor of sympathy and reciprocity?

In the end, whether we consider the conflict of feelings that slavery induces between master and slave or the conflict of feelings it causes within the master himself, we find that at best Darwinism can only report our natural ambivalence with regard to slavery without giving us any compelling reason to either choose or reject it. Arnhart's moral realism, applied to other issues, suggests a different approach to the question of slavery than he offers here. Arnhart notes that marriage is universal and arises from a natural desire for conjugal bonding but that divorce also is universal and arises from a "natural human inclination to serial monogamy."[73] One might say that, in view of the ambivalence of our desires, marriage is, like slavery, naturally unstable. This does not lead Arnhart to declare marriage unnatural, but to affirm the natural status of both marriage and divorce. Consistency would seem to point to a similar judgment of slavery: in light of the ambivalence of our natural desires, both slavery and freedom are in some sense natural. Thus Darwinian natural right, at least if it remains true to its principles, neither requires nor forbids slavery.

CONCLUSION

A moral theory that cannot persuasively condemn slavery cannot, of course, repudiate any less extreme forms of injustice or tyranny, whether perpetrated by majorities, minorities, or individual despotic rulers. The Darwinian account of human nature and morality, then, cannot supply the moral instruction that Tocqueville believes democracy needs. Nor can it, one might add, clearly account for the moral perceptions of most human beings, who seem to discern, despite the intensity of their commitment to those they regard as their own, that something is owed in justice even to those human beings outside their circle of kinship and reciprocal cooperation.

This deficiency of the evolutionary approach, as is now no doubt clear, arises from its reliance on its account of natural feelings as the sole source of moral obligation. One could presumably reject majority tyranny and other forms of inter-group exploitation on the grounds that, though such behavior is tempting in light of our natural tribal desires, it is nonetheless ignoble or unjust and therefore to be restrained. Such appeals, however, are ruled out by the Darwinians, because they imply a source of moral knowledge beyond our evolved feelings. Therefore, in light of the weakness of our feelings for humanity itself, there is not only no natural reason to expect, but also no natural reason to want, politics to be anything more than attempts at exploitation of one group by another. The Darwinian political theorists are correct in their judgment that their science, by demonstrating the natural basis of our sociability and morality, refutes Hobbes's understanding of human nature. But the content of that natural sociability and morality is such that the Darwinian account merely replaces Hobbes's war of "every man against every man" with a war of every tribe against every tribe.[74]

Ultimately, Hobbes offers us more than the war of all against all because he offers us more than nature, which he regards as hostile to human well-being and in need of modification by human ingenuity. He recommends artificial political arrangements calculated by reason to satisfy our most powerful natural passions. Perhaps, then, the Darwinians would respond, adopting such a Hobbesian spirit, as follows. Although our untutored moral feelings provide no basis for universal morality, and therefore no basis on which to repudiate majority tyranny, calculation on the basis of those feelings does. Human beings have evolved not only a complex set of sociable and moral feelings, but also a capacity for rational thought and for action in light of these feelings. As a result, human behavior is much more flexible, much less determined by our elementary passions, than that of other animals. We can, even in the absence of strong feelings of sympathy for human beings as such, respect the interests of outsiders if we see that doing so makes it easier to satisfy our natural feelings to cooperate in our own group. Thus we can see that it is in our enlightened self-interest to respect the rights of others so that they in turn will respect ours.

Nevertheless, this argument presupposes a certain equality between potentially warring groups. It is only in a group's interest to make peace with its competitor if the latter is itself strong enough possibly to succeed at exploiting the former. Under such circumstances the risks to both sides point in the direction of peace. In the absence of such equality, and in the absence of a significant feeling of concern for outsiders, calculation points not toward mutual respect but the tyrannical rule of the stronger. Such equality, moreover, need not exist among competing groups. Indeed, the problem of how to secure protection for rights would not be such a vexed question in political thought had it not often been the case that various groups had succeeded, through disparity of power, in exploiting others.

The Darwinians might respond that modern human beings have solved the problem of the tyranny of the majority without relying on religion by devising clever institutional arrangements that make it difficult for one group to use political power to oppress another. This is the point, after all, of the American Founders' adoption of such institutions as separation of powers, bicameralism, federalism, and the extended republic. Indeed, Tocqueville himself suggested that a system of separation of powers would of itself solve the problem of majority tyranny.

It is not clear, however, that such institutions, ingenious and effective as they may be, can completely remedy the democratic inclination toward majority tyranny. The Darwinian approach, after all, begs the crucial practical question: why, if human moral nature is as evolutionary theory suggests, should we not try to subvert such institutions when we discern that our group can profit from doing so? It is not sufficient to respond that these institutions are designed to resist such attempts at subversion. This is true, and these institutions have proved remarkably durable. That this durability relies to some extent on favorable cultural conditions—perhaps including a religiously informed commitment to fundamental law as a safeguard of the rights of minorities—is suggested by the adoption and subsequent failure of such American institutions in other countries. And even Tocqueville indicates that those who think modern society can rely on self-interest alone are mistaken.

While Darwinism can provide an account of the biological basis of feelings of sympathy and reciprocity, its account of human nature tends to confine those moral feelings to a relatively small circle of the human race. Because the new Darwinian political science seeks to base moral obligation on our evolved feelings alone, it cannot provide a convincing justification for universal morality and therefore offers little to restrain the tyranny of the majority. On the contrary, on the evolutionary understanding inter-group exploitation appears to have nature's sanction as much as, if not more than, inter-group justice. Once again, from the standpoint of Tocqueville's concerns, modern science cannot replace religion as democracy's source of moral instruction.

5

Darwinism and the Family

HOWEVER WELCOME one might find the Darwinian refutation of relativism, the account of human nature and morality offered by the new Darwinian political theorists is not equal to the requirements of Tocqueville's political science and therefore cannot replace religion's role in that political science. For the morality that can be established on the natural feelings identified by evolutionary science cannot extend very high or very far. That is, these feelings, limited by their origins in genetic self-interest and evolution within small groups, cannot sustain a noble morality that transcends the decent materialism of democracy or a universal morality that recognizes even minimal obligations owed to human beings as human beings.

The proponents of the new Darwinian political theory might respond to these criticisms by jettisoning their concerns with noble and universal morality and emphasizing instead the humbler, yet still crucial, contribution to public morality that their theory can offer. Perhaps they would admit, for the sake of argument, that their account of human nature is incapable of supporting human greatness or opposing majority tyranny, but contend nevertheless that this incapacity need not be regarded as a serious defect. After all, they might argue, we can survive, and even flourish in a way, without moral nobility. A society of orderly and decent hedonists may not be particularly beautiful, but it is at least tolerable, generating great prosperity and according to individuals ample freedom to pursue their own concerns. Indeed, even some conservatives have praised contemporary American society precisely on these grounds. When, for example, President Clinton was entangled in the controversy that led finally to his impeachment, many conservatives were dismayed at ordinary Americans' lack of outrage at Clinton's conduct. Other conservatives, however, presented this public indifference as a kind of virtue, a combination of easygoingness and respect for privacy that supports the enviable freedom Americans enjoy.

The defenders of Darwinian political theory might argue further that evolutionary morality's inability to support respect for universal morality, and its consequent inability to offer a principled condemnation of majority tyranny, is, while perhaps theoretically interesting, practically irrelevant, because universal morality, or respect for the rights and dignity of individuals as human beings, is under no serious threat. As Fukuyama points out, universal morality appears to be here to stay, at least among the residents of the developed world. The "advanced societies of North America and Europe" possess sufficient "hierarchical authority" with which to "enforce universalistic principles of individual rights and citizenship," and therefore it is difficult to imagine such societies degenerating into mere aggregations of "hostile, self-regarding caves or 'burbclaves' where the radius of trust extends no further than the edge of the neighborhood."[1]

Indeed, even Tocqueville, while clearly concerned about democracy's capacity for majority tyranny, concedes that democracy is likely to give rise to a kind of universal sympathy that seems to preclude majority tyranny, or at least very ruthless forms of it. The rigid system of rank in an aristocracy creates groups of human beings with very different opinions and interests, whose ways of "thinking" and of "feeling" are so different that members of one class can "scarcely believe themselves to be a part of the same humanity" as members of another. In contrast, democracy, by destroying rank and class, renders human beings more or less the same and therefore easily able to enter imaginatively into the sufferings of others. The result is that democratic men display "a general compassion for all members of the human species."[2]

On this line of argument, the concerns I have emphasized up to this point are not that pressing. Majority tyranny does not seem to have materialized, and we can live (perhaps even happily) with democratic hedonism. There is, however, a pressing need to restore respect for the ordinary decency of reciprocal cooperation. As Fukuyama notes, while large scale order persists apparently without difficulty, the "great disruption" has eroded small scale order considerably. America and the other developed nations have "lost much of the ordinary morality" that is typically respected within small social groups, and therefore these societies must try to restore "habits of honesty" and "reciprocity."[3] The Darwinian account of human nature demonstrates the natural status of our sociable and cooperative inclinations. It thus allows the defender of the "ordinary virtues" to point out that a life that neglects them will be less than fully satisfying.

Moreover, the Darwinian political theorists could point to their approach's ability to establish the natural status of that institution in which the ordinary virtues are first learned: the family. Many scholars, and probably most citizens, now seem to agree that the decline in ordinary morality is related to the disruption of traditional family arrangements. Darwinism promises to breathe new life into our commitment to the family by offering a scientific basis for a conservative view of the family as rooted in human nature. In addition, the Darwinian political theorists might point out, here the evolutionary science of human nature presents itself not as indifferent to Tocqueville's concerns but as supportive of them. For the insights of Darwinian biology seem to confirm Tocqueville's understanding of traditional family relationships as based upon nature and as essential to the well-being of the larger society.

THE FAMILY AND SOCIETY

A number of the Darwinian political theorists suggest that the family is an essential teacher of the ordinary virtues that make for a decent society. As we have seen, they contend that the social virtues, though rooted in our nature, nevertheless require some habituation if they are to come to their full development. Not surprisingly, such habituation takes place first, and perhaps most crucially, in the family.

Fukuyama's *The Great Disruption* is concerned primarily with the decline, in developed nations, in "social capital," which Fukuyama defines as "a set of informal values or norms shared among members of a group that permits cooperation among them."[4] Although he notes that excessive commitment to kin can have negative social consequences, as when a nepotistic "familism" promotes cooperation within families but discourages it among them, Fukuyama also stresses the family as a source of social order and social cooperation. "Families in the first instance constitute the most basic cooperative social unit, one in which mothers and fathers need to work together to create, socialize, and educate children."[5] And while family cooperation is rooted in biological kinship and looks immediately to reproduction, the habits of cooperation that families foster tend to give rise to institutions and relationships that reach beyond the family and provide benefits to the larger society. Fukuyama notes for example that "small businesses, most of them run by families, account for as much as 20 percent of private sector employment in the American economy and are critical incubators of new technologies and business practices."[6] He concludes his discussion of the family by affirming it as "both a source for and a transmitter of social capital."

James Q. Wilson suggests that the habits that foster "upright" and "decent" behavior are, while "founded in nature," nonetheless "developed in the family."[7] Surveying a body of social science on the relationship of family to the character formation of children, Wilson concludes that "the positive relationship between secure attachment" to one's parents, on the one hand, and "sociable behavior" in later years, on the other, "seems universal." Those "children with the weakest bonds to their parents," Wilson notes, tend to "avoid people or to react to the distress of others with fear, anger, or hostility," while "securely attached children show greater empathy than do avoidant children, probably because, having experienced empathy themselves, they have a greater capacity to show it toward others."[8] As these passages indicate, the habits of sympathy and reciprocity that one learns in the family seem to enable one to reach out to those beyond the family. Wilson concludes his discussion by describing the family as "a continuous locus of reciprocal obligations that constitute an unending school for moral instruction" and noting that we ultimately "learn to cope with the people of this world" because we first "learn to cope with the members of our own family."[9]

Finally, Larry Arnhart adopts Aristotle's view that "all social cooperation ultimately arises as an extension of the natural impulses to sexual coupling and parental care of the young." While some animals give little care to their offspring, the "more social and intelligent animals," such as human beings, offer "parental care" of "great duration and intensity," providing not only food and protection but also instruction in "the habits and knowledge required for living in groups with complex social structures."[10] Here, Arnhart notes, Aristotle agrees with Darwin, who similarly held that the "natural bonding of parents and children is the foundation of all social bonding."[11] And this position is confirmed by contemporary ethology, which has "shown the importance of parental care, especially for primates, as the root of all social bonding."[12]

In recent years, however, the developed societies of the west have seen a considerable weakening of the traditional family. In America and Europe rates of fertility have decreased as women have chosen to have fewer children, rates of illegitimacy have increased as more children have been born outside of marriage, and marriage has declined as more people have chosen cohabitation in preference to marriage and as greater numbers of those who have chosen marriage have ended up getting divorced. This decline in the family has been accompanied, moreover, by a diminution in what Fukuyama calls "social capital," manifested in increasing rates of crime and a diminishing sense of trust, and perhaps more subtly in the general decline in civility of which both conservatives and liberals have complained. [13]

It is not unreasonable to surmise that there is more here than mere correlation, that in fact the disruption of the family has contributed to the general decline in sociability. As the Darwinians point out, the family is where our sociability and morality are first developed, through the habituation and example provided by our parents. Yet the family trends mentioned above all arguably point to a diminution in the family's capacity to provide children with such socialization. Phenomena such as illegitimacy and divorce often result in one parent, usually the mother, shouldering most of the responsibility for raising the child. At best, this means that the child will suffer a general decrease in parental attention: one parent will be absent and the other, lacking the first's assistance with practical matters like earning money and keeping house, will have less time to spend with the child. At worst, it means that the child will have received a socialization in anti-social behavior from the dereliction of familial duty displayed by the absent parent. It is true that illegitimacy need not result in single-parenthood, since many couples with children cohabit instead of marrying. Nevertheless, this solution seems incomplete and unstable, since cohabitation seems to be even more prone to dissolution than marriage.[14] Even within stable marriages, the evidence points to a decrease in the family's capacity to socialize children. The decline in fertility suggests that children are no longer thought to be as desirable as they were in the past, and it would not be surprising if this outlook manifested itself in more subtle ways as well, such as a decreased amount of attention paid to and time spent with one's children. As James Q. Wilson points out, where, as in modern societies, children have less economic value, fewer will be born, and "those children who are produced will be raised, at the margin, in ways that reflect their higher opportunity cost. Some will be neglected and others will be cared for in ways that minimize the parental cost in personal freedom, extra income, or career opportunities." Indeed, this point need not be left to speculation: Wilson points to evidence that parental investment of time with children has declined in America since the 1960s.[15]

What accounts for this diminishing commitment to the family? To some extent the decline can be traced to changes in material conditions. Modern economic circumstances, as Wilson suggests, render children less economically useful. In an economy organized around agriculture, children do not necessarily need a sophisticated and expensive education and, in any case, can defray to some extent the cost of their own rearing by contributing their labor to the family's agricultural efforts. In modern economies, however, parents must pay for a costly education to maximize their children's chances of success in life, and there is little that the children, while they are still children, can contribute to the family economy.[16] In addition, Fukuyama points to the role of modern technology in diminishing modern man's attachment to the family. On the one hand, the birth control pill, by liberating women from the possibility of unwanted children, liberates men from a sense of responsibility for the children they father and for their mothers: after all, they reason, the woman was free to prevent conception had she really wanted to. On the other hand, modern technology has economically devalued sheer bodily strength and increased the value of intellectual capacity, thus removing the economic advantage that men had over women when labor was predominantly physical. As a result, women are better able to support themselves through their own labor and their economic need for marriage has accordingly diminished.[17]

The decline of the family may also be due in part to the influence of ideas. Fukuyama points out that since the 1960s, in both America and Europe, "the liberation of the individual from unnecessary and stifling social constraints" has been a "very powerful cultural theme," and a number of "liberation movements" have emerged to "free individuals" from the dominance of "many traditional social norms and moral rules." As examples of such movements Fukuyama points to the "sexual revolution, the women's liberation and feminist movements, and the movements in favor of gay and lesbian rights."[18] Thus his argument implicitly, and correctly, indicates that the traditional family itself, as well as the norms surrounding and supporting it, is one of the primary social constraints from which many in the west have sought liberation.

One of the leading intellectual tools of such liberation, however, is the denigration of the idea that the family is a natural institution and that the moral obligations that attend it are rooted in human nature. If, for example, sex does not belong by nature within marriage, if the old-fashioned notion that this is the case was merely an irrational cultural taboo, then one may seek sexual pleasures with any partner of the opposite, or even the same, sex. If one has no duty to one's children that is by nature prior to one's desire for intellectually stimulating work or contributing to the larger community, then one may, without guilt, pursue a career or public service in preference to the exercise of parental obligations. The popular dismissal of the natural status of the family and of traditional sexual morality has been supported, moreover, by much contemporary social science which, as was noted in chapter 2, tends to deny that there are any moral obligations or social structures rooted in human nature, and which therefore suggests that there is an endless variety of ways in which human beings may permissibly and wholesomely live.

If the family is essential to the formation of character with a view to a decent sociability, on the one hand, yet has been undermined by modern economics, technology, and ideology, on the other, modern society evidently stands in need of some teaching that can restore commitment to the family. In other words, we need some account of the family as natural, so that we can once again see its obligations as sources of fulfillment and not merely as constraints, and so that we can appreciate the need to respect it even in modern circumstances that allow us easily to abandon it. Such a teaching is offered by the new Darwinian political theory, which reveals the roots of the family in our biological nature and which therefore can present it as responsive to our natural desires.

THE NATURAL STATUS OF THE FAMILY

In defense of the natural status of the family, understood as a social arrangement in which parents care for their own biological children, Larry Arnhart points to the extremely problematic attempts, undertaken by various socialist communities, to do away with the parent-child relationship. Generally such social experiments have failed outright, or at least led to considerable dissatisfaction on the part of their own members. This suggests that the bond between parents and offspring is not merely a cultural institution that can be modified or eliminated through new educational and social arrangements but rather is based on deeply rooted natural desires. Arnhart surveys the history of the Israeli kibbutzim, secular socialist communities that have sought, among other things, to raise children communally. As early as two weeks after birth infants were placed in communal nurseries, after which their "entire care and education would be communal." Older children continued to live in public "dormitories under the care of trained nurses and teachers," and the young were not "permitted to sleep in their parents' apartments."[19] While such arrangements worked initially, they gradually gave way to a restoration of more traditional, parent-provided child rearing —at the insistence of the parents themselves. Mothers, in particular, demanded a time during the day when they could visit their children and wanted to put them to bed at night. Eventually, Arnhart notes, such "small deviations from communal norms" developed "into a major revival of family autonomy," such that now "the private family centered on parent-child attachments has become the basic unit for rearing children" in most of the kibbutzim.[20] Moreover, that such backsliding is evidence of natural familial desires, and not merely the result of the persistence of culturally learned norms, is suggested by the fact that the leaders of the movement to restore the private family were not members of the founding generation, who had been raised in private families prior to the establishment of the kibbutzim, but instead those who had themselves been raised on the kibbutz and who therefore should have been accustomed to its social arrangements.[21]

Arnhart further notes that not even the invocation of accepted religious authority can render such a community of children palatable to most human beings. In the nineteenth century, he points out, John Humphrey Noyes and his followers founded the Oneida Community in an effort to practice "Bible Communism" and establish on Earth the "we-spirit" of Heaven. As on the kibbutzim, children were raised communally. Separated from their mothers at weaning, children were placed in a separate house and then allowed to visit them only two hours a week—unless they showed signs of excessive attachment, in which case "even those brief visits would be denied."[22] Despite assurances that they were, by submitting to such social institutions, participating in the establishment of heaven on earth, members of the Oneida Community often fell prey to "the 'sin' of 'partiality' or 'special love,'" which was most commonly manifested as "the attachment of mothers to their children."[23] As a result of this and other difficulties the community was finally disbanded. Arnhart notes that such experiences point to the natural status of the family. While the history of the Oneida Community demonstrates that the private family can be abandoned for a time, it also shows that "this will be felt as a *sacrifice*," even by those raised in such a community, who one would expect to be unproblematically habituated to such arrangements if the private family were merely a cultural construct. "The attachment between parent and child," Arnhart concludes, "is too important to the natural organization of human desires to be suppressed without creating an emotional cost that will be unendurable for most people."[24]

It is not surprising that such natural familial proclivities should exist, moreover, because they obviously serve our natural biological interests. Insofar as all organisms seek to reproduce their genes in the next and subsequent generations, and insofar as human offspring require intensive care if they are to survive and flourish, it would be strange indeed if parents did not normally care for their children and find it emotionally satisfying to do so. "If we accept Darwin's theory of evolution by natural selection, which favors those functional adaptations that promoted survival and reproduction in evolutionary history," Arnhart observes, "it would seem likely, therefore, that the desire to care for children is a natural adaptation for human beings."[25] Similarly, James Q. Wilson contends that "family processes" do not depend on education so much as on "instincts" that "were forged by millennia of natural selection." Those who "would not care for offspring," or who did so poorly, would not have left many "surviving offspring," and the natural result is that "the family is a universal human institution."[26]

Once again, however, this is not to say that the Darwinian account requires us to view parental care in purely utilitarian terms, as if parents treat their children as means to the end of passing on their genes. Evolution selects for not so much a generalized desire to survive and reproduce as for a complex set of desires that promote those natural ends. Yet these desires are themselves experienced as ends by the organisms in which they have evolved. Parents "do not care for their children because they have consciously calculated that this promotes reproductive fitness" but because they have a powerful and "natural desire" to do so.[27] A mother does not feed and play with her baby because she wants it to survive and pass on her genes, she does it because she wants to, because it makes her happy—or at least, in view of the admitted difficulties that attend child-rearing, it makes her happier than if she left them undone or left them to someone else.[28] To say that parental care responds to a powerful natural desire is to say, however, that parental care is obligatory, at least for anyone who wants to lead a satisfied and happy life. Thus Arnhart concludes that "[p]arents caring for their children is a fundamental norm of Darwinian natural right."[29]

The Darwinian account of parental care thus appears to provide a solid foundation for addressing the difficulty noted earlier: the decline of the family and the consequent decline in society's capacity to socialize children in decent and cooperative behavior. Contrary to what one might be tempted to conclude from the family's decline under modern economic and technological conditions, and contrary to what many liberationists might want to believe, the family is neither a mere function of variable social and material conditions nor simply a product of cultural learning. It is a manifestation of our nature, and consequently the obligations it imposes are not simply artificial constraints but also sources of natural fulfillment and happiness. Thus the Darwinian presentation of human nature can offer a persuasive account of why people should be willing to undertake the difficult work of socializing children, upon which the decent and cooperative behavior of future adults depends.

This Darwinian account of the family, moreover, seems generally to correspond to Tocqueville's understanding of the family and the contribution it makes to society. In Volume I of *Democracy in America*, Tocqueville suggests that respect for the family is the cause of America's relatively secure social order, a security which tends to elude Europe because of its failure to adhere to the natural obligations of family life. "In Europe," Tocqueville contends, "almost all the disorders of society are born around the domestic hearth" and "not far from the nuptial bed." As a result of their experience in the European family, men acquire a "scorn for natural bonds and permitted pleasures," a "taste for disorder," a "restiveness of heart," and an "instability of desires" that render them problematic as members of the larger community and because of which they submit "only with difficulty" to the public authority. In contrast, Americans tend spontaneously to support public order because they encounter "order and peace," and "simple and natural" pleasures, in their households. "While the European seeks to escape his domestic sorrows by troubling society," Tocqueville sums up, "the American draws from his home the love of order, which he afterwards brings into affairs of state."[30]

Tocqueville's argument, then, indicates that scorn for the natural bonds of the family creates citizens who are only weakly attached to society as a whole and that, on the other hand, the careful cultivation of those natural obligations ultimately forms characters that are attached not only to their own kin but to the well-being of the wider community. Tocqueville thus suggests, like the Darwinian political theorists, that familial relationships are rooted in nature, that they are the basis of social bonding and cooperation beyond the family, and that their neglect tends to produce human beings whose capacity for sociable interaction is impaired.

THE NATURE OF TRADITIONAL SEX ROLES

Tocqueville, however, contends not only that the obligations of family members are rooted in nature, but also that those obligations are naturally different. That is, he affirms the natural status of traditional sex roles of husbands and wives, fathers and mothers, within the family. In Volume II of *Democracy in America* he takes up the American understanding of the "Equality of Man and Woman." Noting that democracy generally tends to destroy or at least modify "the various inequalities to which society gives birth," Tocqueville wonders whether it will also ultimately "act upon the great inequality of man and woman," which had until recently seemed not merely a product of social arrangements but instead to "have its eternal foundations in nature." He concludes that democracy "elevates woman and must, more and more, make her the equal of man," but he quickly suggests that here a distinction is necessary, lest one conclude that the equality of the sexes requires that men and women fulfill the same familial and social functions.[31]

There are, Tocqueville points out, those in Europe who, under the influence of democracy's egalitarian ideology, confuse "the diverse attributes of the sexes" and who seek "to make man and woman into beings not only equal, but alike," imposing on "both the same functions" and "the same duties." Such sexual egalitarianism, however, goes against nature's grain and therefore "degrades" both men and women. By denying them habituation in the activities for which they have a real aptitude, and by simultaneously seeking to habituate them in activities for which their temperaments are not suited, such an approach impedes the development of the distinct, though equally valuable, excellences of which the respective sexes are capable. From the "coarse mixture of nature's works" that European egalitarians attempt, Tocqueville insists, "only weak men and disreputable women can ever emerge."[32]

In contrast, Tocqueville argues, the Americans have not understood the equality of the sexes in this manner. In America, Tocqueville finds, men and women are assigned, and expected to fulfill, distinct social functions. One does not find "American women directing the external affairs of the family, conducting a business, or indeed entering the political sphere," which activities are apparently reserved to men.[33] Rather, American women tend to the internal needs of the family, and in particular, one supposes, to the rearing of children. Thus Tocqueville observes that women in the United States are confined by public opinion "within the small circle of interests and domestic duties," and that American women themselves agree that for them "the sources of happiness are within the conjugal dwelling."[34]

Such social arrangements are, unlike those to which European egalitarians aspire, in conformity with nature, that is, with human nature or the respective natures of men and women. The Americans, Tocqueville observes, "have thought that since nature had established such great variation between the physical and moral constitution of man and that of woman, its clearly indicated goal was to give a diverse employment to their different faculties."[35] Our modern opinion is that the natural differences between men and women amount only to differences in reproductive organs and that any differences in temperament must be the result of social learning. Against this view, Tocqueville suggests that these physical differences are accompanied by naturally different "moral" or psychological inclinations that generally suit men and women for different social tasks. Tocqueville does not here specify the relevant differences of faculties, but he surely has in mind the traditional notion that men are more aggressive and therefore better suited to conducting the family's external relations and to the rough-and-tumble world of politics, while women are more nurturing and therefore better equipped to tend to the care and education of children.

Because its social and familial arrangements are in conformity with nature, moreover, Tocqueville's America gets better results from them than it otherwise would. By assigning to the sexes those tasks for which they have a natural aptitude, the Americans ensure that those tasks are performed in an excellent manner. With nurturing women managing the family's internal affairs, children receive the attention and solicitude that they need; and with competitive men conducting the family's external affairs, outsiders are handled with the detachment that is necessary to protect the family's interests. Tocqueville remarks that the "Americans have applied to the two sexes the great principle of political economy that dominates industry in our day. They have carefully divided the functions of man and woman in order that the great work of society be better done."[36]

The Darwinian account of human nature appears to support Tocqueville's traditionally conservative understanding of the proper sexual division of labor. Arnhart notes that much contemporary opinion, especially that of feminist social theorists, holds that "the differences in the typical social behavior of men and women are not natural but purely cultural," that the apparent "differences in the psychological dispositions of men and women" are not "natural" but instead only the result of "social learning" that is grounded on "arbitrary cultural traditions." In response to such notions, Arnhart contends that the sexes "are different by nature," not only physically but also psychologically: "The natural pattern of desires typical for men is not the same as that typical for women." Among other differences, men generally "have a stronger desire for dominance, while women typically have a stronger desire for nurturance." These different natural desires, moreover, point to different natural excellences of character: "the distinctive virtue of men is courage, while the distinctive virtue of women is sympathy."[37] Such differences, Arnhart concedes, are "typical or average" and therefore not to be observed in all comparisons of particular men and women. Some women (say, Margaret Thatcher) are more dominant than many men. Nevertheless, Arnhart's account suggests that these natural disparities are sufficiently widespread and deeply rooted that they are relevant to our thinking about how to organize our social life.

As evidence of the natural status of these differences one can point, once again, to the experiences of communities that have tried to do away with them. Both the Oneida Community and the Israeli kibbutzim sought to dispense not only with the private rearing of children, but also with traditional sex roles that were thought to be oppressive to women. In both cases, however, psychological and behavioral differences between the sexes proved as difficult to eliminate as did the attachment of parent to child. As noted before, in the Oneida Community the most common offenders against the prohibition on partiality were *mothers*; and on the kibbutzim the leading critics of the communal child-rearing were *women* who wanted to spend more time with their biological children. Such developments tend to confirm Arnhart's assertion that the desire for nurturance is more powerful in women than in men. In addition, as both Arnhart and Wilson point out, social scientists who studied the kibbutzim over a long period of time, and who began with the assumption that psychological differences between the sexes were purely the products of social conditioning, came away convinced that the persistence of such differences pointed to their rootedness in nature. As Arnhart observes, although the "environment of social learning in the kibbutz had been designed to discourage" behavioral differences between boys and girls, researchers observed that during play girls were more likely than boys to pretend to be mothers, while the boys displayed "more conflict and aggressive behavior."[38] Wilson notes that the boys on the kibbutz were "more likely to pretend they were driving machines" while the girls were "more likely to play with dolls and baby buggies." As they grew, the "formal sexual equality of life continued," but the manifestation of sexual differences persisted: "Girls began assisting the nurses, boys the farmers. Teachers began to remark that girls were socially more sensitive, the boys more egotistical." When disputes arose, girls tended to offer "assistance, sharing, and cooperation, while the boys more often relied on initiating activities, applying rules, or issuing directives. Predictably, the most aggressive children were boys. Both boys and girls would attempt to control the aggression, but only the girls would console the victims of it."[39]

The natural status of such differences is further indicated by their apparent basis in the physiology of the brain. Arnhart notes the existence of "sexual differences in the central nervous system of mammals" that likely contribute "in complex ways to sexual differences in social behavior." Although these differences are not as pronounced in primates as in other, less sophisticated mammals, they are present, and presumably present in humans as well.[40] Such differences in behavioral propensities appear to be rooted in the natural, biological interests of male and females. As has been suggested before, those interests, primarily in reproductive fitness, are to some extent the same for all human beings, and therefore the Darwinian political theorists can point to the existence of a common pattern of human desires, evolved in the service of our biological interests, that can be the basis of a common, natural human morality.

On the other hand, it is not the case that males and females manifest their interest in reproductive fitness in identical ways: their reproductive strategies tend to be different, and it is therefore not surprising that they should have evolved distinctive sets of natural desires. Darwinian biologists observe that females by nature invest a good deal more in any single instance of reproduction than do males. While the latter need only fertilize, the former must gestate and then nurse the offspring. As a result of these differing levels of investment, males can, in principle, produce far more offspring than females. The female has a great deal more of her potential reproductive fitness riding on the success of any given child than does the male, consequently has a natural interest in investing a good deal of care in each child, and therefore has developed a set of natural desires that incline her to nurturing behavior.[41]

The Darwinian account of human nature, then, appears to lend support to Tocqueville's argument that men and women have different temperaments by nature and therefore that society is better served when they apply themselves to distinct tasks. Tocqueville, however, also insists on a kind of equality between the sexes. He observes that democracy tends to place different members of society more and more on the same level, and while he is concerned, as we have seen, that this progress of equality not obliterate the traditional sexual division of labor, he appears happily to embrace a sexual equality that stops short of this and respects the natural differences in aptitude between men and women. This explains his praise for the esteem Americans show for women and for the social tasks assigned to them. While the "Americans do not believe that man and woman have the duty or the right to do the same things," they nonetheless "show the same esteem for the role of each of them, and they consider them as beings whose value is equal although their destiny differs." While they have "allowed the inferiority of woman to subsist in society," insofar as women are assigned to nurturing rather than governing, they have nonetheless "elevated her with all their power to the level of man in the intellectual and moral world," insofar as they see to the education of women and respect their intellectual capacity and social functions. Thus have the Americans "admirably understood the true notion of democratic progress."[42] Here again, the Darwinians can affirm a similar position, because there is nothing about the nature of the differences between men and women that suggests that one sex is superior to the other. On the contrary, the qualities they display, like nurturance and competition, are, though divergent, both clearly necessary to the flourishing of society. Accordingly, Arnhart and Wilson alike dismiss the notion that one of the sexes is more moral than the other.[43]

MARITAL FIDELITY AND SOCIAL STABILITY

Tocqueville also affirms another kind of sexual equality, one that he praises the Americans for respecting and blames the Europeans for disregarding: equality in adherence to strict standards of marital fidelity. Europeans, he observes, seem to have given the man "a sort of singular immunity" in these matters," such that "there is almost one virtue for his use and another for that of his mate." According to European "public opinion, the same act can be alternatively a crime or only a fault." In contrast, "Americans do not know this iniquitous division of duties and rights," and therefore "[a]mong them the seducer is as dishonored as his victim."[44] Moreover, that the Americans resolve this inequality in the direction of regarding adultery uniformly as a "crime" rather than a "fault," and that Tocqueville approves of such moral rigor, is suggested by Tocqueville's argument elsewhere in *Democracy in America*. He observes that Americans are uniquely intolerant of sexual irregularity, and he suggests that the "great severity of mores that one remarks in the United States" is a sign that America is the country "where the bond of marriage is most respected and where they have conceived the highest and most just idea of conjugal happiness."[45] Although Tocqueville nowhere expressly approves the permanence of the marital union, such approval is surely implicit in his praise of the American understanding generally: divorce was hardly looked upon with the same indulgence in 1830s America as it is today.

It is at this point that Tocqueville and the Darwinian political theorists begin to diverge. For, despite its ability to justify the family as a natural institution, the Darwinian account of human nature cannot sustain the lofty conception of marital fidelity endorsed by Tocqueville. As noted before, on the Darwinian account, men and women share a common good: reproductive fitness. The nature of that good, however, does not appear to require a permanent union, and consequently does not foster the evolution of the kind of feelings that would sustain permanence of marital commitment. Moreover, the very different ways in which men and women pursue and experience that common good seems to undermine the possibility of a high standard of mutual commitment even while a marriage lasts.

There is little reason to suppose that the biological good at which the conjugal union aims would require parents to remain together longer than is necessary to raise children to an age at which they no longer require intensive parental care. Humans are drawn to familial commitment by the unique, and uniquely lasting, dependence of human children. Because human offspring mature at a comparatively slow pace, their parents must remain together for some considerable time if they hope to reproduce their genes in the next generation. On the other hand, once the children have progressed sufficiently, there would remain no natural basis for the continuation of the relationship of the parents. Arnhart observes that divorce is a universal phenomenon and that humans "tend to form pair bonds that last long enough to rear their young during the years of infancy when children are most in need of care by both parents."[46] Human beings, he suggests, possess a "natural" propensity to "serial monogamy," which moves them to "marry, divorce, and then remarry."[47] Although Arnhart admits that couples may "develop ties of stable attachment that can last for the rest of their lives," there appears to be no sense in which it would be an important departure from nature if this failed to happen.[48]

Indeed, to the extent that the independence of children will tend to be accompanied by the natural beginnings of infertility in the mother, a man would, on Darwinian grounds, have a natural incentive to leave the mother and seek a new, younger partner capable of bearing more children. Such behavior serves male biological interests and is encouraged by the innate desires that have been selected to advance those interests. A number of evolutionary theorists have observed that men tend to be attracted to physical beauty and youthfulness in females because these traits are good, although not infallible, indicators of fertility. Of course, in terms of such characteristics the mother of mature children will usually suffer by comparison to many other women, and her husband will therefore experience a powerful, and natural, desire to move on to someone else. In addition, men will have ample opportunities to act on this desire. Because the female is primarily preoccupied with nurturing the few offspring she can produce, she is attracted less to youth and beauty than to status and wealth, which are signs of a man's ability to invest resources in her children.[49] Therefore, the father of mature children, who will have had more time than a younger man to move up the social hierarchy and acquire wealth, will likely appear much more attractive to a young woman than a younger man. He will certainly have more appeal to the opposite sex than the wife who has spent her youth rearing his children. The male behavior to which these circumstances point must be regarded as perfectly natural, on Darwinian grounds. Fukuyama observes that it is not surprising that "in most societies . . . wealthy or high-status men have greater sexual access to women," which, in societies that forbid polygamy, often manifests itself as "corporate executives," for example, having "their wives and children serially rather than simultaneously."[50]

One may say, moreover, that, on the assumptions of the new Darwinian political theory, this account of human nature not only fails to give support to notions of the permanence of marriage, but actually fosters hostility toward such notions. The Darwinian political theorists seek to derive moral standards solely from our evolved human nature, from the biologically rooted desires that have developed in support of our reproductive interests. Yet the understanding of marriage enforced by Tocqueville's Americans seems to make demands on human beings that go beyond, and indeed run contrary to, our natural desires. Hence Arnhart's condemnation of the Catholic understanding of the indissolubility of marriage as "an imprudent dogmatism that goes against human nature."[51]

If the Darwinian account of human nature does not support the notion of permanent marital commitment, neither does it point to a very strict standard of mutual commitment while a marriage lasts. While there is a common good shared by spouses, the reproductive success they seek to advance through their offspring, that good is itself perceived, pursued, and valued by men and women in very different ways. Again, and as Fukuyama points out, a woman invests much more in any single child than a man will precisely because she can produce so few offspring in comparison to the man: "For humans, a woman can produce perhaps a dozen children in a lifetime, while a man can sire thousands."[52] In light of the limited number of opportunities they have to reproduce, women have a strong incentive, and hence have developed a desire, to seek committed relationships with men who can provide resources to support their children. Men, on the other hand, have a powerful biological interest in, and hence have evolved a desire for, promiscuity: they "tend to maximize their chances for passing on their genes by mating less discriminatingly with as many females as possible."[53]

Because of these remarkably different biological aims and desires, the union of man and woman in the service of reproduction will frequently involve a good deal of manipulation and deception. Thus Fukuyama observes that "the same sex act tends to be interpreted differently by men and women . . . Even if both end up having sex, their intentions are different, and one or the other usually ends up having been deceived."[54] He develops this point later in *The Great*

Disruption, and appears to suggest that a natural basis for a common good between men and women is almost completely lacking. "Males and females," he notes, "are constantly playing games with each other, males seeking to maximize the number of their sexual partners and females seeking the fittest male who can provide for them and their offspring." Men, therefore, have powerful interests to "pretend" they "will provide resources and loyalty when" they have "no intention of doing so," while women have "strong incentives to detect this deception." In contrast, women have "a strong incentive" to ensure that their "children are fathered by a male with the best possible genes, regardless of whether he is" her husband or provider, while men have "a strong incentive to avoid being cuckolded and wasting" their "resources raising someone else's offspring."[55]

The evolutionary account of human nature, then, does not point to a very rigorous standard of marital fidelity. Men will have a natural inclination, arising from their reproductive interest in sowing the seed as widely as possible, to sexual promiscuity that will lead them to desire sexual relations with women other than their wives. Women, because of the need to maximize the chances of success of each child they bear, will be inclined to sexual relations with men other than their husbands if they think that their adulterous partners have superior genetic endowments. On the other hand, as Fukuyama points out, each partner has an incentive and an accompanying desire to prevent such betrayals. It would seem that the Darwinian marriage would not be very stable.

Indeed, despite the incentive to stability provided by the need to rear young children, the evolved desires of human beings would in many cases spur a dissolution of marriage even during the period when children still need intense care. On the Darwinian understanding, a woman, and especially an attractive and young woman, might well be inclined to leave her husband if she had an opportunity to take up with a man with superior resources who was willing to support the children she already had. More problematic with a view to family stability are the more promiscuous desires of men. While concern with the well being of his children deters a man from abandoning his current partner for someone new, it is not clear that such a concern would be a naturally more powerful desire than his longing for sexual variety, which can in some cases be satisfied by leaving his family. As both Francis Fukuyama and James Q. Wilson point out, while maternal care for children is deeply rooted in female biology, the role of men in providing parental care appears to be much more a product of social construction.[56] This weakness of paternal care arises, again, from the male capacity to generate so many offspring, in light of which it is much less risky for a man simply to father children and then move on, leaving the extent and quality of their care beyond his control.

Larry Arnhart, however, seeks to overcome these difficulties, and thus to provide some evolutionary basis for the stability of marriage, by contending that male and female natures can achieve a kind of harmony. While the differences between masculine and feminine reproductive strategies and sexual inclinations are a source of tension, there is nevertheless "a fundamental complementarity" between the desires of men and women that "supports the stable arrangements of family life."[57] Men admittedly are powerfully inclined to promiscuity, but they also have, like women, "deeper desires for conjugal stability and parental care." Marriage, therefore, can be understood as a fulfillment of male as well as female human nature. After all, Arnhart concludes, if "marriage did not satisfy male desires, it would not be a universal practice of all societies."[58]

While this argument establishes marriage and the family as natural, it does not seem sufficient to support a very strict standard of marital fidelity. Even if one grants Arnhart's assertion that the male desire for conjugal stability and parental care is by nature deeper and more enduring that the male desire for the promiscuous pursuit of sexual pleasure, one would still have to admit that the latter inclination is itself very powerful and enduring. Thus, on a theory that seeks to derive morality solely from our evolved desires, the best arrangement, for men at any rate, would be to combine marriage with such philandering as does not unduly disturb the marital bond—perhaps by finding a mate submissive and dependent enough to tolerate it or by choosing occasions that minimize the chances of detection. Perhaps this is why Arnhart is careful to note that "men have desires for intimacy and security that are frustrated when they *yield completely* to their desire for sexual variety."[59]

The Darwinian account of human nature, then, provides little support for Tocqueville's notions of marriage as demanding permanence and exclusivity of commitment. This is not to say that the Darwinian account of the typical sexual desires of men and women is false. On the contrary, any candid reader will admit that it rings true. The difficulty, rather, is that the new Darwinian political theory, because it seeks to ground its morality solely in our evolved desires, cannot offer an account of marriage that is any more demanding than the typical desires of ordinary human beings. For these Darwinians, morality is derived from what is naturally desirable for human beings. Given this premise, and given the common pattern of male desire that they observe, these theorists would have to hold that adultery and even abandonment of the family are, for many men, not only understandable natural temptations, but even naturally good. After all, such behavior, while it will no doubt carry a certain cost in guilt, may in many instances result in the satisfaction of more, or more powerful, desires than it frustrates. And in that case there would be no basis, on the argument of the Darwinian political theorists, on which such temptations should be resisted.

THE VIRTUES OF THE TRADITIONAL FAMILY

In defense of the Darwinian account one might ask why such temptations should be resisted. That is, this critique of the evolutionary account of marriage is only valid on the assumption of the superiority of Tocqueville's more demanding account. But why should we embrace that assumption? After all, and as Arnhart's remarks about Catholic teaching on marriage suggest, the Darwinian account indicates that Tocqueville's demands for permanence and exclusivity in marriage go well beyond our natural desires, contradict our nature, and therefore frustrate our natural happiness.

In response to this objection we can point to at least three reasons to prefer Tocqueville's more demanding understanding of marital duty. In the first place, while adherence to Tocqueville's standards may admittedly be difficult to the extent that it requires us to discipline powerful natural desires, for precisely that reason those standards provide the occasion for the display of a noble self-control and a generous, because irreversible, commitment of loyalty to another person. Tocqueville contends that such generosity of mores, manifested in a willingness to subordinate one's desires out of commitment to another, is more common in aristocracy because of its insistence on the inviolable bonds of feudal fealty. In a democracy, however, where human relationships are organized to a much greater extent around the material self-interest of the parties, such generosity is comparatively rare. In this light, the traditional understanding of marriage that Tocqueville defends may be, in our modern democratic circumstances, the only remaining occasion for the display of a noble loyalty.[60] Thus the abandonment of this understanding represents a moral loss. The Darwinian account of human nature seems to necessitate such a loss, however, because it must view permanence and exclusivity of marital commitment not as difficult but obligatory, but as difficult and unreasonable.

Second, we may observe that the costs of abandoning these loftier notions of marital fidelity appear to fall more heavily upon women than men. The propensity for promiscuity is more pronounced in men, and women appear spontaneously to place more value on commitment than do men. The obvious consequence of surrendering the old notion that adultery is flatly impermissible would be—in fact, has been—the liberation of men to pursue extramarital liaisons. Their wives will still feel these liaisons are betrayals but will have little interest in pursuing liaisons for themselves as a compensation. Moreover, as was suggested earlier, embracing the impermanence of marriage means in practice that many men will leave their wives for younger women, while the wives themselves will have very limited prospects for obtaining a new spouse. As Fukuyama notes, "[f]or reasons that are directly biological (men remain more sexually attractive at later ages than do women)," abandoned first wives have "much lower chances of ever remarrying than the husbands" who leave them." It is in the light of such considerations that Fukuyama contends that one of the "greatest frauds perpetrated during the Great Disruption was the notion that the sexual revolution was gender neutral, benefiting women and men equally."[61]

Finally, and most important in the context of the concerns relevant to this chapter, the more demanding conception of marriage seems necessary to achieving the social stability that both Tocqueville and the Darwinians seek. As noted before, Tocqueville remarks, in Volume 1 of *Democracy in America*, the Europeans' disrespect for "natural bonds" and their consequent experience of social disorder. These observations, however, immediately follow his praise for the "great severity" of American mores as manifesting "the highest and most just idea of conjugal happiness." Apparently, the "tumultuous passions" that disturb European families and European societies are adulterous passions, and the American conception of marriage is not merely admirable but essential to the well-being of the larger society. Hence Tocqueville's comment that in Europe "all the disorders of society are born" not only "around the domestic hearth" but also "not far from the nuptial bed."[62] Tocqueville restates these concerns in Volume II's discussion of democracy and sexual morality. "In the eyes of the legislator," he suggests, "prostitution is much less to be feared than intrigue." While the former brings about the "great corruption of some" and therefore produces "deplorable individual miseries," the latter involves "the laxity of all," and this "puts society in danger" by threatening to "destroy family bonds" and to "enervate national mores."[63]

It is reasonable to suppose that Tocqueville is correct in linking sexual irregularity with social instability. To the extent that a society tolerates adultery it fosters in men a generalized uncertainty about the paternity of their children. A man living in such a society will, all other things being equal, be less likely to invest as much attention or effort in the support of his wife or the socialization of his children. He will in fact be more likely simply to leave the family and thus completely withdraw from his role in their socialization: there is the real possibility that the children in his care are not actually his offspring. Even where paternity is certain, the tolerance of adultery combined with male sexual jealousy will induce many men to regard their own wives as possible or actual betrayers. Such a society will be destabilized not only eventually—by the behavior of the children whose character formation will have been harmed by the weak commitment of men to their families—but also immediately—by the unruly actions of men with only a weak commitment to their children and wives. Because of the weakness of that commitment, such men have no strong stake in the present and future stability of the society at large and are therefore more willing to disturb its tranquility in the pursuit of their own immediate interests. Indeed, in a culture that tolerates adultery a man is more likely to view the surrounding society as composed chiefly of enemies. Rather than seeing other citizens as fellow enjoyers of the goods of family life, he will regard them as either potential aggressors who seek to seduce his wife and destroy his domestic happiness or as potential dupes whose wives he can seduce. Hence Tocqueville's observation that "the European seeks to escape his domestic sorrows by troubling society," while "the American draws from his home the love of order, which he afterward brings into affairs of state."[64]

As was noted earlier, Tocqueville says nothing explicit about the importance of the permanence of the marital commitment, no doubt because this principle was almost universally acknowledged when Tocqueville wrote. Nevertheless, one can discern that permanence, as well as exclusivity, contributes to social stability by contributing to the positive socialization of the young. That socialization in decent and cooperative behavior results in part from children's observation of the interaction of their parents, of how well they treat each other. Surely, however, there would be at least a subtle decline generally in the decency with which spouses treat each other when a marriage is undertaken and carried out on the assumption of its impermanence, when it is known all along by both partners that they need not habituate themselves too much in caring for the other because the time may soon come when they need not care for each other at all. James Q. Wilson concludes his treatment of the family, as we noted earlier, by suggesting that we "learn to cope with the people of this world because we learn to cope with the members of our family."[65] His argument, which he supports with a quotation from Chesterton's *Heretics*, implies that we learn to cope with the members of our family because we have to. But how well will we learn to cope with them, to bear with their weaknesses and foibles, when we know that we need not cope with them for long? Moreover, once it is accepted that a marriage may come to an end, it is likely that one of the partners, probably the man, will want to end it sooner than the other, which will likely lead to considerable and ugly conflict in the family—a conflict that will become an inescapable and undesirable part of the child's socialization.

The Darwinian account of the family and society confronts us with the following dilemma: the natural desires of parents point to one understanding of marital commitment, while the full moral flourishing of children and the needs of society point to another, more demanding, understanding. More specifically, one might say that the natural desires of men lend themselves to a weak commitment to the family that is contrary to the well-being of women and children, and to the long-term interests of the larger society. The Darwinian political theorists are aware of this difficulty, and they consequently point out that if the family is to flourish, the natural desires of males must be supplemented, or rather disciplined, by social learning.

James Q. Wilson notes that sexual selection will militate to some extent against male promiscuity and detachment from children. That is, since women will try to choose as mates men who seem more inclined to remain faithful and to assist in caring for children, men who manifest such qualities may be more likely to achieve reproductive success, which means such qualities will likely be rooted to some extent in male nature.[66] On the other hand, the aforementioned logic of male reproductive interests will constantly limit the extent to which sexual selection can succeed in producing these qualities. Indeed, as Fukuyama suggests, males have a biological interest in preventing the female's accurate identification of such traits. Thus one male response to female sexual selection is to *appear* to have the desired fidelity and concern for children while secretly harboring more promiscuous desires.

As a result, nature, and particularly male nature, is insufficient as a basis for the family. Wilson begins his account of sex differences by asserting that "[n]ature has played a cruel trick on humankind" by making "males essential for reproduction but next to useless for nurturance" and simultaneously making such nurturance essential if children are to "grow up in an orderly and safe environment and be part of an elaborate and useful culture."[67] Wilson accordingly concludes that female sex selection "is not strong enough to produce families unless it is powerfully reinforced by cultural expectations and social sanctions."[68] Francis Fukuyama likewise finds that the male role in the family is by nature "fragile and subject to disruption" and therefore that the "extent to which males will stay in monogamous pair-bonds and play an active role in the nurturance of children will depend less on instinct than on the kinds of social norms, sanctions, and pressures that are brought to bear on them by the larger community."[69] Even Larry Arnhart, who, as we have seen, stresses the ways in which the family satisfies male longings, notes that our natural mating desires are difficult to satisfy because they tend to contradict each other. He hints that social learning is required to establish family stability. Marriage as a "social, economic, and sexual bond between husband and wife," he suggests, is not simply spontaneous but the result of social prescription.[70]

This natural dilemma and its evident social solution raise the following question: to what authority will society appeal in teaching men to curb their powerful promiscuous propensities for the sake of family and society? On the teaching of the Darwinian political theorists, which claims that a sufficient public morality can be established on the basis of our natural desires alone, society will have to appeal to the desires of men as they are revealed by evolutionary science. But these must be desires other than the natural male inclination to familial commitment that is admittedly too weak to sustain the family. Fortunately, natural selection has given rise to some other male desires or interests that would seek to vindicate the family. Because men's reproductive interests would not be advanced if they were tricked into providing resources to children who are not their own offspring, men have a natural desire, sexual jealousy, to prevent their own mates from mating with others. In addition, to the extent that a man may be a father or brother of an adult woman, whose reproductive fitness is to some extent his as well, he will have an interest in securing her husband's commitment to her and her children.

Unfortunately, however, such natural interests will always exist in any given man along side his own powerful inclination toward promiscuity. It seems, therefore, that the entire set of male desires will be ambivalent with regard to commitment to family. While men will want their own wives and other women's husbands to be faithful, for themselves they will want to stray to the extent that they can get away with it. On this view, society's commitment to marriage will be, for men at least, a kind of truce among sexually competitive men. But as Glaucon indicates in Plato's *Republic*, where justice is considered merely a tolerable mean between what is best, doing injustice without suffering it, and what is worst, suffering it without being able to do it, those who can succeed in doing what is "best" on this understanding will have no incentive not to do it. The Darwinian account, then, can provide no unequivocal reason that a man should not merely pay lip service to fidelity and insist on it for others, while practicing sexual adventurism for himself.

Tocqueville can provide such a reason. The "great severity of mores" that he observed, which accompanied America's possession of "the highest and most just idea of conjugal happiness," had its "primary source," Tocqueville contends, "in beliefs," in the religion the Americans professed.[71] Tocqueville immediately emphasizes, however, that the influence of religion upon American men is to a considerable extent indirect. While it often fails to restrain men, he says, "it reigns as a sovereign over the soul of woman, and it is woman who makes mores."[72] Christianity, by teaching that sexual irregularity is a serious sin, makes women much less likely to cooperate with the promiscuous desires of men, either by tolerating adultery in their husbands, if they are married, or by engaging in adultery with married men, if they are unmarried. By making women as a class generally intolerant of adultery, Christianity makes exclusive and permanent commitment to a wife the price of male sexual satisfaction. While such a solution may not necessarily reform the male character, it at least is sufficient to provide for familial stability and the flourishing of society.

One need not conclude, however, that on Tocqueville's account American men embrace the Christian understanding of marriage only hypocritically and out of necessity. As we noted earlier, in Tocqueville's America women, coordinating their social duties with their natural inclinations, devote themselves primarily to the management of the internal needs of the household and, presumably, to the rearing of children. It is therefore reasonable to suppose that such pious women shape not only the mores of the current generation of men, through their own sexual conduct, but also, through the moral education they provide their children, the mores of the next generation of men. In Tocqueville's America, we may hope, boys will internalize their mothers' Christian commitment to marital fidelity and thus grow into men who want to control their unruly appetites. Hence Tocqueville's suggestion, noted earlier, that American men really are happy with their family arrangements.

Where nature alone lacks the capacity to vindicate the family, it seems, God's command is necessary. Yet the stability of the family is, as the Darwinian political theorists admit, essential to the well-being of society itself. Therefore, religion appears necessary to the proper functioning of even the most apparently natural human institution, and even to the minimal decency and sociability to which that institution so powerfully contributes.

6

The Demise of
Darwinian Morality

THUS FAR, our argument has suggested that the Darwinian refutation of relativism, while sound as far as it goes, cannot sustain certain moral principles that democracy requires and that, in some cases, even the theorists of evolutionary ethics want to affirm. The argument up to now has implicitly conceded at least this much: that the Darwinian account of human nature can serve as the basis of some morality, inadequate though it may be to modern democracy's moral needs. In the end, however, Darwinism cannot sustain any recognizable morality at all, and this ultimate failure arises not so much from the relative poverty of the feelings it is able to justify as natural, but from the very attempt to treat feelings as the sole source of moral knowledge.

THE MORALITY OF FEELINGS

The new Darwinian political and moral theory seeks to base its moral teaching on feelings, on what it takes to be the natural human desires or passions. This point is made most clear by Larry Arnhart and Robert McShea. "We are," McShea contends, "the sum of our feelings." Therefore, our "only motivation to doing is feeling," and the only "conceivable good for us" is the "maximum satisfaction of our strongest and most enduring feelings."[1] Or, as Arnhart sums up, "The good is the desirable."[2]

The fact-value distinction is commonly posited as the rational basis of moral relativism. Science or reason is thought to analyze facts, but, as David Hume argued and as value-neutral social scientists have repeated many times, it is a logical fallacy to derive a statement about what ought to be from a statement about what is. Because of the intellect's inability to affirm any values as superior to others, human morality is consigned to the realm of arbitrary preference.

The proponents of evolutionary morality, however, contend that our natural desires are the basis for an overcoming of the fact-value distinction. "Description," McShea suggests, "leads to prescription" by the mediation of desire.[3] Indeed, Arnhart contends that Hume has been widely misunderstood on this issue and that he in fact intended a derivation of morality from feelings similar to that offered by the Darwinian account.[4] To say that an "ought" cannot be derived from an "is" does not mean that morality is non-existent, only that it is grounded in human passion rather than speculative reason.[5]

Seeking to be consistent with this understanding, the Darwinian political theorists are careful to reject abstract reason as the basis of moral knowledge. Arnhart, for example, repudiates Kant's notion of "morality as an autonomous realm of human experience governed by its own internal logic with no reference to anything in human nature such as natural desires or interests."[6] McShea likewise rejects "God," "Nature", and "Reason," understood as transcendent principles, as the basis of morality.[7] Values are not, he insists, "'out there,' somehow embedded in the structure of things external."[8] Similarly, James Q. Wilson endorses Hume's rejection of reason alone as a basis of obligation and pointedly remarks that he prefers to say that morality rests on "sentiments" rather than "intuitions"—presumably because the latter term suggests not a feeling but a direct intellection of moral truth by the rational mind.[9]

One might think that this rejection of reason and embrace of feelings would lead directly to, rather than save us from, moral relativism. To recall C.S. Lewis's argument in *The Abolition of Man*, if one holds that "all sentences containing a predicate of value are statements about the emotional state of the speaker," will one not naturally draw the conclusion "that all such statements are unimportant," that "all values are subjective and trivial"?[10] The Darwinians, however, deny that this conclusion follows. According to them, while moral judgments are not purely rational and cannot be understood to have a universal objectivity that transcends human nature itself, neither are they merely subjective or idiosyncratic reflections of individual emotions. Following Hume, Arnhart contends that "[m]oral judgments do not have *cosmic objectivity* in the sense of conforming to structures that exist totally independent of human beings," but neither do they "have only *emotive subjectivity* in the sense of expressing purely personal feelings."[11] Rather, morality arises from feelings that are universal among human beings because rooted in their evolved nature. Precisely by being widely shared, these feelings are a basis of moral agreement. Thus moral judgments achieve a kind of "*intersubjective objectivity* in that they are factual judgments about the species-typical pattern of moral sentiments in specified circumstances."[12]

Similarly, McShea argues that morality is not merely a question of one's point of view because "moral judgments are composed of two variables, circumstances and feelings," both of which are facts that can be known. Therefore, for "a given set of known feelings embodied in an individual or a group, in a given set of known circumstances, there are humanly objective and in principle knowable better and worse moral decisions."[13] Like Arnhart, McShea characterizes his moral theory as "intersubjective," noting that moral communication typically involves not a mere assertion of one's feelings but a kind of calculation, on the basis of our shared humanity, that others will experience similar feelings. McShea points out that when "someone says 'Brown is hateful,' what is usually meant is 'I hate Brown, and if you knew certain facts about Brown you too would experience feelings about him that would lead you to say that you hate him also.'"[14]

Thus the Darwinians reject Lewis's denigration of values arising from emotions. The emotions from which moral judgments arise, they contend, are the basis of our distinctive species-nature and therefore are both important and common rather than trivial and subjective. Reason's powerlessness to establish values independently does not mean that there is no moral truth. Rather, a "true" moral judgment is one that corresponds to the "sentiment" or feeling of "normal human beings under standard circumstances."[15]

THE MORALITY OF PSYCHOPATHY

If morality arises from the feelings of normal human beings, what, one wonders, is the moral status of the feelings, judgments, and actions of abnormal human beings? Confronting this question leads to a remarkable paradox: the Darwinian account of morality, which claims to provide a way out of moral relativism by providing a scientific basis for moral judgments, ends up ruling out the one moral judgment that most human beings would probably find most obvious, namely, that the psychopath is evil. From this paradox, as we will see, follows the complete unraveling of the Darwinian claim to support any morality whatsoever.

While constructing their ethical theory on the basis of the normal feelings of ordinary human beings, the Darwinian political theorists also recognize the existence of abnormal humans who experience, and sometimes act on, feelings that are not ordinary and not, at least from the perspective of the ordinary person, in any sense moral.

Arnhart, for example, while contending that "there is a natural moral sense rooted in human biology," admits that "some people, if only a few, seem to have no moral sense." Categorized as "psychopaths" by "modern psychology," such human beings are "ruthless social predators" who, because they are "incapable of feeling the pain of others," manipulate other human beings for their own ends, which sometimes amount to no more than the satisfaction of a mere desire for excitement, "with no feelings of guilt or regret."[16] Similarly, while noting the slight feeling variations that exist among normally constituted human beings, McShea concedes the existence "feeling deviants," those who, through either natural mutation and variation or the influence of "extraordinary circumstances," fail to experience certain normal feelings or experience them in an unusual manner.[17] James Q. Wilson notes that there are some human beings who, while possessing logical minds, nevertheless are "emotional cipher[s]," "nonsocial" personalities "for whom the ordinary emotions of life have no meaning" and who therefore can "lie without compunction, injure without remorse, and cheat with little fear of detection."[18]

It is not the mere existence of psychopaths that poses a problem for Darwinian moral theory, however, but the non-judgmental posture toward psychopathy that the Darwinian moralists are compelled to take by their reliance on feelings alone as the ground of morality. Because moral obligation arises from feelings and cannot possibly arise from anything else, it follows that psychopaths are not immoral to have or even to act on their abnormal feelings. Arnhart makes this clear. "If morality depends upon a moral sense," he argues, "and if the moral sense depends on moral emotions that typically arise in most human beings, then it would seem that those few people who happen not to share those moral emotions are not bound to obey that moral sense." Psychopaths, therefore, are "neither good nor evil," are "under no moral obligation to conform to the moral sense, because they lack the moral emotions that provide the only basis for moral obligation."[19] McShea likewise contends that, because "values follow upon feelings," the "commonality of species feelings has no normative authority over those" few human beings "who do not share it." For example, although the overwhelming majority of humans "condemn child abuse, you are not wrong if you feel that it is enjoyable. If you do not share this part of the normal feeling pattern, you are not bound by the morality that depends on it."[20] "Human feeling deviances," as McShea sums up elsewhere, "cannot be considered bad in themselves" but only bad "in relation to our feelings."[21]

One might well assume that such an argument results in a kind of moral disarmament diametrically opposed to what the Darwinian theorists explicitly hope to achieve: in James Q. Wilson's words, the recovery of the "confidence" with which ordinary people "once spoke about virtue and morality," the recent loss of which "may have abetted, by excusing, the few who have no moral sense."[22] Arnhart and McShea, however, hasten to deny this consequence, contending that just because the psychopath is not wrong it does not follow that society is wrong to control him. On the contrary, because their emotional constitution is such that there can be no fruitful moral communication with psychopaths, our relationship to them must be mediated entirely by coercion. Lacking the moral sense, Arnhart argues, "psychopaths must be treated by the rest of us as moral strangers whose dangerous conduct can only be restrained by force." While they "are not morally obligated to respect social desires that we feel but they do not, neither are the rest of us morally obligated to refrain from protecting ourselves against their predatory behavior."[23] Similarly, having just given child abusers a green light from the standpoint of their own internal feelings, McShea immediately offers an amber light of caution based on the external consideration of the feelings of most other human beings. The child abuser, he notes, "would be wise to suppress" any "display" of his "species-idiosyncratic feelings." "The species-human feeling for children is very strong," he notes, and if the abuser is "not wrong in not sharing it" the rest of us are also not wrong "in taking whatever action against" the abuser is "appropriate to safeguard the children."[24] Just as, in the absence of any other source of values, the psychopath may act on his feelings, so we may act on ours, on the feelings of fear and repugnance that the psychopath inspires in normally constituted people.

Maintaining this position leads McShea, at least, into a thicket of absurdities. In his "Introduction," McShea decries moral relativism, under the influence of which, he rightly notes, we can say nothing to the "child abuser, the racist, the terrorist" except that "we do not like what they do" and that "if we are stronger than they we will make their practices costly to them." Nevertheless, we remain, McShea suggests, "uneasily aware that our likes and dislikes, the accident of our superior force, are not arguments at all."[25] He presents his account of human nature as the answer to this intolerable situation, yet, as we have just seen, McShea has in the end nothing better to offer than the relativism he criticizes. His response to the child abuser is nothing but an appeal to feelings of normal humans, or the likes and dislikes of most of us, and to our superior force, our ominous ability to do whatever is necessary to protect our children.

In his refutation of "culturalism," McShea ably brings to light the incoherence into which those are forced who contend that culture is the only basis of value. On the culturalist presupposition, McShea points out, the ways of a culture cannot be criticized from within, because the culture itself and its ways are the sole source of morality. Neither can they be judged from without, on the basis of the ways of some other culture, for essentially the same reason. Such a position, however, is evidently self-contradictory. If, McShea points out, our own culture condemns certain actions, and yet we see members of other cultures performing those actions, then from culturalism "it follows that" we "ought not condemn them, for the values of" our "culture do not apply to theirs," but at the same time that we "ought

to condemn them, for the values of" our "own culture are by definition imperative upon" us.[26] McShea's effort to establish morality on
feelings alone, however, places the normal human being in exactly the
same problematic position in relation to the psychopath or feeling
deviant: on the one hand one cannot condemn him, because one's
own feelings, which he does not share, are the sole basis of value,
and on the other hand one must condemn him, because one's own
feelings are outraged by his actions.

Sensitive, perhaps, to the difficulties that arise when one main-
tains that the psychopath is bad only in relation to our own feelings,
Arnhart makes some effort to demonstrate that psychopathy is also
objectively bad, at least in the sense of being unsatisfactory even to
the psychopath. The psychopath, he contends, has a "flat soul" that
is largely unmoved by any of the normal human emotions, including
a prudent concern for his own well-being.[27] The title of Arnhart's
chapter on this topic is "The Poverty of Psychopathic Desire." Because of this emotional poverty, moreover, psychopaths cannot lead
a successful life. "The same emotional poverty that prevents their
caring about the feelings of others also prevents their caring about
their own future," with the result that they are incapable of guiding
"their behavior to any deliberate goal" or according to "any consistent
purpose." Though their craving for excitement and indifference to
others might seem to suit them well for professional crime or ter-
rorism, Arnhart concludes that, because even these activities require
self-control in the pursuit of some deliberately chosen aim, psychopaths would inevitably fail here as well.[28]

Arnhart's critique of the psychopathic life, however, is unper-
suasive—at least on the principles he himself articulates. Arnhart's
critique is sound in itself, but it presupposes a capacity to stand outside of one's feelings that his own theory insists is impermissible. It
makes no sense to speak of poverty of feelings unless one postulates
that there are certain feelings that one ought to have; but if feelings

themselves are the source of value, this cannot be the case. Is a pig impoverished because of his inability to enjoy the works of Bach? Not on Arnhart's understanding. Because such enjoyment is no part of the pig's natural repertoire of feelings and behaviors, he cannot be considered impaired because of its absence. Similarly, the psychopath cannot experience as "emotional poverty" a lack of sociable or moral feelings that he does not possess and has never really experienced. Nor can one reasonably say, on the basis of the ethics of feelings, that a psychopath cannot be successful. The impossibility of success, Arnhart contends, stems from the psychopath's lack of ordinary self-love. Yet if this is the case he will not experience his life as a failure. A man who has no strong feelings of concern for his own well-being cannot find his own lack of well-being bad in any important way.

Despite such self-contradictory efforts to avoid it, the Darwinian conclusion with regard to psychopathy amounts to this: psychopaths are not bad, but just different. Such a teaching seems strange coming from a theory that hopes to restore the credibility of the value judgments of ordinary citizens.

THE MACHIAVELLIAN PERSONALITY

One might offer the following defense of the Darwinian political theorists. The problems arising from their treatment of psychopathy, though theoretically interesting, are of no practical relevance. Psychopaths are very rare, and therefore the failure adequately to account for their moral status presents no serious difficulty. A moral theory need only speak to the ordinary range of human experience.

What Arnhart terms the Machiavellian personality, however, is more common. Machiavellianism, Arnhart suggests, following such scholars as Linda Mealy and Robert Hare, is a kind of "low-level" or "subcriminal" manifestation of psychopathy.[29] Presumably, the Machiavellian, like the ideal prince described by Machiavelli himself, is characterized not by the complete absence of self- and other-regarding feelings, and the consequent psychopathic impulsiveness, but instead by a limited, but perhaps ultimately more dangerous, participation in them. That is, he cares enough about himself to seek worldly success, but not enough to eschew high-risk strategies in its pursuit; and he cares enough about others to desire lofty positions attended by public esteem, but not enough to give the interests of others anything like the weight he accords to his own. Arnhart suggests, again following Hare, that Machiavellians "are just as selfishly manipulative" as pure psychopaths but still possess sufficient discipline "to get what they want while appearing to be morally normal."[30]

As this remark suggests, the Machiavellian's life, unlike the psychopath's, cannot so easily be described as a failure. Arnhart concedes that Machiavellians "can sometimes attain worldly success," noting that they are "effective social manipulators" because of their ability to "pursue rational strategies for winning every social competition without being distracted by an emotional" or "moral commitment to people or principles."[31] Nor can the Darwinian account dismiss such human beings as aberrant. On the contrary, Arnhart admits that Machiavellianism may to some extent be a natural phenomenon. "[E]ven in a population that is predominantly cooperative," he points out, "it would be advantageous for a few individuals to adopt a cheater strategy so that they could exploit the cooperative behavior of the general population."[32] There may be an "evolutionary niche for Machiavellians," and, because of their capacity for success, "natural selection might favor some traits of the Machiavellian temperament."[33]

Here again, the difficulty is not the mere existence of Machiavellians, but the evaluation of them that Darwinian moral theorists like Arnhart and McShea must adopt—an evaluation that, like their evaluation of psychopathy, reveals their theory's alienation from our ordinary moral experience. Presumably, Machiavellianism is wrong.

Certainly most people would regard as deplorable the manipulation of others with a view to one's own social success, without, or with only minimal, feelings of regret. Yet Darwinian morality has nothing to say to the Machiavellian. In acting amorally, he is doing right by his own feelings, which is all that anyone can be expected to do. If he finds power and glory more attractive than friendship and pity, then for him those things *are* better, because feelings are the sole ground of value judgments. We can try to find ways to deter him, but no moral appeal can be made to him.

Nor can the Darwinian teaching say anything morally useful to the potential Machiavellian. The existence of such persons is suggested by the Darwinians' proper observation that nurture completes nature and therefore that morality, though natural, develops to some extent through habituation. Arnhart remarks that "in pursuing some narrowly selfish desires" we might fail "to cultivate those social bonds of affection and cooperation that we need to satisfy our social desires."[34] McShea similarly observes that the "ordinary human feelings" can be "desensitized" at "any age."[35]

The potential Machiavellian, then, is faced with a choice. He can try to cultivate his moral feelings and restrain his Machiavellian desires, or he can commit to the satisfaction of the latter and therefore to the steady desensitization of the former. Which should he do? The Darwinian moral theory can give him no guidance. One might point out that choosing Machiavellianism will close off certain pleasures to him, but if the feelings that lead to those pleasures can be desensitized, then their absence need not in the end be experienced as a loss or deficiency. Besides, the Machiavellian can point in response to the pleasures that come with glory and power, which cannot be securely enjoyed without an ability to manipulate, without sympathy, people and situations.

The difficulties posed by the Darwinian account's reliance on feelings alone become even more acutely practical when we realize that all human beings are to some extent potential Machiavellians. Evolution, Fukuyama indicates, will not (and obviously has not) tended to produce a race of angels or utterly selfless virtuous beings. Rather, it produces societies of "people who have different proportions of angelic and demonic qualities at the same time."[36] Human nature, it seems, is a mixture of moral and non-moral inclinations. This makes sense on the evolutionary explanation of human nature. While it may be true that some amount of cooperative behavior is conducive to genetic fitness, it is no less true that some amount of selfish or amoral behavior is as well. And as we have evolved to experience some measure of moral behavior as pleasant, we have evolved to experience many selfish or amoral behaviors as pleasant as well. Roger Masters indicates that "human behavior is a compound of contradictory impulses."[37] It is safe to assume, he argues, that among primates possessing complex systems of socialization natural selection will favor "both selfish and cooperative behaviors." Human nature, therefore, displays "a complex of cooperation and competition."[38]

Ultimately, on this view, our moral proclivities, though real, are nevertheless no more real than our amoral proclivities, with which they may indeed be inextricably intertwined. Masters, following Robert Trivers, remarks that "humans seem to specialize," on the one hand, "in moralistic aggression against those who cheat," and equally, on the other, in "deceit to cheat on their partners" and "deceitful imitation of compliance and morality to avoid punishment."[39] According to this understanding, while we may say with confidence that human beings are by nature moral creatures, we may not say that they are *fundamentally* moral creatures. Thus it is not clear why our social or moral feelings should be accorded any more exalted status in our lives than our selfish or amoral ones.

This is not, by the way, to embrace what Arnhart, following Frans de Waal, rejects as "Calvinist sociobiology," which adopts what Arnhart takes to be the dualism of the Augustinian and Kantian accounts of morality, holding that nature is evil and that ethical conduct therefore requires a complete repudiation of our humanity.[40] It is to say that we have a mixed nature, a combination of impulses, and that Darwinism can give us no good reason to prefer the things commonly taken to be good over those commonly taken to be bad. This incapacity arises, again, from Darwinism's insistence on basing morality on the feelings of our evolved nature with no reference to any other principles. If our evolved nature is the basis of morality, then its darker elements are just as legitimate as its lighter ones. If the desirable is the good and feelings the sole source of value, then all desires are moral and all feelings provide their own ethical justification.

RIGHT HAS MIGHT?

In the face of this difficulty the various proponents of Darwinian morality react in various ways. Some simply appear unaware of the problem, speaking conventionally of the "good" and "bad" sides of our nature, apparently forgetting that on the evolutionary account human nature is itself the sole ground of value judgments. This tendency is most pronounced in *The Moral Sense*, in which James Q. Wilson speaks of "our better instincts" in contrast to "our baser ones," even while admitting that the former may be "fainter" than the latter—as if there were some standard of knowledge of better and worse beyond our natural feelings.[41]

In contrast, Arnhart and McShea at least confront the difficulty directly, seeking to justify the moral feelings as more choiceworthy than the non-moral ones by positing that the former are ultimately more powerful than the latter. Nothing else, it seems, will serve to establish the superiority of the moral impulses if one insists on the exclusion of any appeal to anything but the feelings. Arnhart therefore indicates that the sociable and moral desires are "more important or

enduring than others."[42] McShea likewise contends that "there are
no good or bad feelings," but that some nevertheless have more of a
claim on us because of their superior "strength." For example, there
is "nothing good about parental feeling in itself," but "we approve of
it because it is a strong and enduring feeling in almost all of us."[43]
Finally, then, McShea suggests that "natural morality" is based upon
"the strongest and most enduring feelings."[44]

But why should we accept this easy solution? Given the stability
of social order, and the ordinary decency of most human beings, it is
certainly reasonable to think that we are naturally moral. On the other
hand, given the necessity of the threat of punishment to sustaining
that order and fostering that decency, and the commonplaceness of
crime and other forms of less aggressive immorality despite such
punishments, it is certainly reasonable to doubt that the moral desires
are decisively more powerful than the immoral.

CONCLUSION

Darwinism clearly succeeds in showing that morality—or at least part
of what we commonly call morality, the ingredients of elementary
decency, such as family feeling, sympathy, and reciprocity—is rooted
in human nature. To this extent it is certainly preferable to cultural
relativism, with its insistence on morality's utter artificiality. Darwinian moral theory, however, taking its understanding of evolved
human nature as its only standard, cannot show, indeed is forced to
deny, that the moral elements of our nature are any more natural, any
more fundamental, than the other elements. It cannot show why we
should prefer them over the others. It can, of course, point out that,
because the good is the satisfaction of desires, the happiness of most
people depends on their attentiveness to their moral sentiments; but
then it cannot deny that full happiness might also require no less
attentiveness to their non-moral sentiments.

Finally, then, the Darwinian account seems to point to some opportunistic balancing of moral and non-moral impulses as the best way of life for the ordinary person. This is certainly better than an outright rejection of morality, but it is not a principled affirmation of a moral life, of the obligation to treat fellow citizens well, which is what most people ordinarily mean by morality. The Darwinian political theorists finally offer only a "morality" that cannot coherently prefer normality to psychopathy, cannot prefer principled decency to opportunistic Machiavellianism. Such an approach clearly fails to reaffirm the moral judgments of ordinary people or to satisfy the aims of the Darwinian political theorists themselves.

7

The Abolition of Man

T HE CONCERN animating Tocqueville's analysis of democracy, and the aim of his new political science, is, as we have seen, the preservation of the conditions of human freedom and dignity. To him this seemed a daunting task. The modern forces of dehumanization and despotism, such as obsession with material comforts and unreflective submission to the authority of the majority, arise naturally from the democratic social state, while the principles capable of sustaining resistance to such democratic evils, such as religious belief, are themselves undermined by democracy. It is little wonder, then, that even as he praised democracy for being superior in justice to aristocracy, Tocqueville still expressed a feeling of dread at the ominous progress of equality.

As grim as modern man's prospects seemed to Tocqueville, however, his concerns might now seem quaint in comparison to the new threats to human freedom and dignity posed by modern technology, and in particular by the prospect, now no longer merely speculative, of the wholesale manipulation of human nature through biotechnology. Such bioengineering threatens human freedom because of the (until now) unheard-of power that it places in the hands of those who will wield these new technologies. The progress of technology is commonly regarded as the "conquest of nature." Yet, as C.S. Lewis points out in *The Abolition of Man*, because all people do not ordinarily have equal access to the power conferred by new technologies, the power of humanity generally cannot be said to be increased in any simple way by progress in technological capacity. Indeed, on careful consideration "what we call Man's power over Nature turns out to be a power exercised by some men over other men with nature as its instrument."[1]

In the case of the new biomedicine, the power of technology will be wielded by parents and scientists and applied to children. The progress of science is beginning to make it possible for parents to exercise unprecedented control over their offspring. As our knowledge of the relationship of brain chemistry to mood and behavior increases, neuropharmacology will enable parents chemically to alter their children's emotions and actions. Indeed, as Francis Fukuyama points out in *Our Posthuman Future*, the popularity of the drug Ritalin indicates that this power is already coming into being. Cloning, embryo screening, and genetic engineering will create the possibility of an even more radical parental power over offspring, insofar as such technologies will allow parents to actually determine a child's genetic design, his biological nature.

Of course, parents have always sought to impose their will on children and have always tried to modify the behavior of the young through instruction, rewards, and punishments. In the past, however, such character formation was limited by the given material, that is, by the child's unique nature. The traditional father may have tried to make his son into what he wanted him to be, but because of the limited tools at his disposal he had also to respect what his son was. The reasonable course to which this tension points, and which prudent parents have adopted, is to teach children to control the problematic manifestations of their underlying character and to encourage them in socially useful, or at least harmless, displays of it. Biotechnology, however, promises to give parents fundamentally new powers: rather than seeking merely to modify how a child lives out his unique character, neuropharmacology will allow parents chemically to switch off aspects of his character that they find troublesome, and cloning and genetic manipulation will allow them to determine that character for him in advance. As Leon Kass points out, such uses of biomedicine—and in particular cloning and genetic engineering —radically alter the relationship between parents and children by creating an unprecedented inequality between them: while natural reproduction implicitly affirms the child as whatever he turns out to be, genetic manipulation seeks to the maximum possible extent to make the child what the parents want, and nothing else.[2]

The new biotechnologies threaten human dignity because they threaten the conditions under which that dignity can be displayed. The great good that modern biomedicine is said to achieve is the relief of suffering. Its proponents constantly trumpet this benefit while showing little awareness of the relationship between human suffering and human dignity, ignoring or ignorant of the opportunity that suffering provides for displaying excellence in patient endurance or creative overcoming. As Aristotle points out, it is in misfortunes that virtue and greatness of soul shine forth all the more. This is not to say that one should actively seek suffering, or refuse opportunities to end it, out of a desire to manifest heroism in living with it. On the contrary, Aristotle's frequent and approving references to the physician's art indicate that he takes ordinary human health, which entails the absence of unusual suffering, as one of the natural goods with which his science of human life is concerned and which human beings ought to seek. Nevertheless, Aristotle's insight into the dependence of certain important goods, like virtue or nobility, on our very limitations and the difficulties with which we must struggle, reminds us that human suffering could not be completely abolished without incurring some loss of our stature as human beings. Thus his approach points to the possibility of distinguishing which human limitations should be remedied through medicine—such as unusual ones that depart from the normal healthy functioning of most human beings—and which should be borne as the essential occasions for the practice of moral virtue.

Because of their implicit presentation of the relief of suffering as the human good, the ideologists and practitioners of the new biotechnology give little reason to hope that they will be able to make such distinctions. Indeed, given the medical establishment's embrace of cosmetic surgery—which seeks not merely to correct the disfiguring maladies or injuries of patients but to provide customers with more shapely noses, chins, or chests—it would seem that the broader culture within which the proponents of the new biomedicine operate either lacks the intellectual resources to recognize, or the moral seriousness to adhere to, such distinctions. It is not surprising, then, that the fledgling efforts of biotechnology have quickly moved beyond therapy to enhancement, to the detriment of moral achievement. While Ritalin can be used legitimately to correct chemical imbalances that lead to genuine Attention Deficit-Hyperactivity Disorder, it has also been used to ease the "suffering" of parents dealing with ordinarily rambunctious children, relieving those parents of an opportunity to show patience in bearing with and creativity in channeling their children's spirited energies. While Prozac can be used legitimately to remedy defects in brain chemistry that lead to pathological depression, it has also been used to cheer up people who are ordinarily sad, relieving them of the opportunity to seek some serious undertaking that might energize and joyfully occupy their souls.

As such uses indicate, and as the example of Prozac vividly illustrates, there is a powerful temptation to use the fruits of biotechnology not only on our children but even on ourselves—again, with a view only to removing irritating difficulties with which we can see no point in struggling. Thus such technology may make possible not only a despotism of some over others that Tocqueville did not foresee, but also a dramatic enhancement of the democratic soul's despotism over itself—more specifically, of the tyranny of the lower over the higher elements of the human soul that democracy fosters. Such biotechnology as now appears on the horizon is simply an extension of the democratic materialism that Tocqueville observed in the early nineteenth century. Its primary aim is to render life constantly easier and more comfortable, to remove all physical and even psychological distress.

It is, as Tocqueville notes, a kind of degradation to submit oneself entirely to such aims, but biotechnology makes possible a submission to them unimaginable to the early American men of commerce and industry. It is one thing to organize one's life, in the sense of one's ordinary activities, around the pursuit of such petty aims. It is something else to alter one's life, in the sense of the biological and chemical basis of one's existence, in the pursuit of them. To put it another way, while it is a kind of degradation for men to devote their minds entirely to such goals, how much more perfect is their dehumanization when they do this and then use the fruits of neuro-pharmacology or genetic engineering to silence the restiveness and boredom that would otherwise alert them to the unworthiness of the goods they so ardently and unhappily seek.

As these reflections indicate, these more recent and more radical threats to human dignity, while no doubt inconceivable to Tocqueville himself, are nonetheless perfectly understandable on the basis of his analysis of democracy. Tocqueville attributes the restiveness of the American spirit in part to its insatiable, and therefore forever unsatisfied, love of equality. When feudal aristocracy "is the law of a society," he contends, "the strongest inequalities do not strike the eye," but "when everything is nearly on a level," as in democracy, "the least of them wound it." Democrats can "easily obtain a certain equality" but can never "attain the equality they desire"; and it is "to these causes that one must attribute the singular melancholy that the inhabitants of democratic lands often display amid their abundance."[3] The same psychology evidently governs equally the democratic taste for physical comfort. Just as even the slightest inequalities seem outrageous once men have become equal in almost all respects, so the most minor causes of unease appear more monstrous as comfort is constantly increased. Hence the observation, made in rare moments of self-recognition, that the residents of the developed nations of the world are spoiled: we cannot stand discomforts that are nothing compared to what peasants of a few generations past endured with patient and even good-humored fortitude. In fact, the unquenchable democratic longing for material comfort can be seen as a manifestation of its insatiable appetite for equality. When every inequality irritates the soul, we naturally view it as an injustice that we suffer any evil or even inconvenience of which others are free.

Ultimately, then, democratic citizens are ill-equipped to make the distinctions necessary to limit the applications of biotechnology because they experience any unsatisfied want as a kind of intolerable suffering. They tend to be blind to the threat biotechnological enhancement poses to human dignity because the goods that make up that dignity—such as the strength of soul that is able to endure suffering and exertion in the pursuit of moral excellence—are considered by democratic men, if they think of them at all, as far less urgent than ease and comfort. Nor is it likely that scientists themselves will impose restraints on the uses of biotechnology. As Tocqueville points out, because the materialistic majority sets the moral tone in democracies, democratic scientists are much more interested in practical applications of their knowledge, that is, applications that serve popular appetites, than in theoretical principles. It is hard to imagine that democratic scientists would seriously entertain philosophic concerns about biotechnology's implications for the meaning of human existence. Given modern science's agnosticism with regard to ultimate questions of purpose and meaning, contemporary scientists will be nothing more than democratic citizens with technical expertise, and hence just as eager as ordinary citizens to apply technology as far as possible to the "relief of man's estate." It is therefore not surprising that modern biomedical science consistently justifies itself, and is consistently embraced by citizens, on the basis of what Nietzsche termed the "religion of pity"—a distinctly modern and democratic religion that cannot stand any kind of suffering and that is therefore hostile to any elevation of human aspiration.[4]

Such are the threats to human dignity and freedom in our time. Any serious conservatism must be able to offer a credible response to these dehumanizing possibilities. Surely a conservatism that cannot explain why we should conserve human nature itself is not worthy of its name. One might well suppose that a conservatism informed by the new Darwinian political theory could offer grounds for resisting the technological reconstruction of our humanity. If, as this theory contends, human nature is the basis of all that we experience as good, then it would seem impermissible to alter it. But as we will see, the Darwinian account of human nature offers no compelling grounds for rejecting, and in fact seems to invite, what Francis Fukuyama has termed "our posthuman future."

THE DARWINIAN DENIAL OF COSMIC TELEOLOGY

One might object to the alteration of human nature through technology by positing that our nature is objectively good and therefore worthy of our respect. Leon Kass suggests something like this when he notes that human cloning represents a radical departure from sexual reproduction, and he condemns cloning as a violation of our "given human nature."[5] Such a formulation, however, raises the obvious question: by who or what is our nature given to us, and what authority does the giver possess to insist that we keep it as given? If our nature is the handiwork of a God far superior to us in benevolence and wisdom, we would, of course, refrain from reinventing it according to our own whims. Alternatively, if our nature were the product of some rational design, not proceeding from the wisdom and goodness of a personal God, but simply rooted in the nature of things, at the basis of the universe itself, then, again, we might well refuse to manipulate it through technology.

But the Darwinian account of the origins of life denies human nature such transcendent bases as could render it objectively good and hence respect for it absolutely obligatory. According to Darwinism, human nature, like all apparent order and design in the realm of living nature, is the result of a mindless process, a combination of chance and necessity. The advocates of the new Darwinian political theory all admit this. Fukuyama, in the context of his discussion of the biological basis of the human moral sense, notes that "throughout the natural world, order is created by the blind, irrational process of evolution and natural selection."[6] That human beings came to be as we know them, Larry Arnhart notes, is, on the Darwinian view, the "contingent" result "of an evolutionary process that could have turned out differently."[7] Roger Masters points out that evolution is not "a progress toward better forms or adaptations" that can be "attributed to a conscious intention or plan." Rather, all organisms are "the product of a trial-and-error process," a competitive "game that populations of organisms play against the rest of nature."[8] James Q. Wilson similarly points to the contingent origins of human nature when he remarks that the organisms "that we see today are the descendents of those that had traits . . . that made them reproductively successful in their *particular* environment."[9] Indeed, according to the evolutionary understanding, not even the reproductive success of organisms is to be viewed as an intention of nature. Rather, it just happens, as a necessity, that "animals that are physically and psychologically put together in such a way as to favor the survival of their genes in direct or collateral descendents are more likely to have such descendents, and those descendents will inherit the favoring characteristics."[10]

In sum, as Arnhart points out, while Darwinism can sustain an "immanent teleology" based on "the striving of each living being" to attain the ends implied by its natural desires, evolutionary theory "surely does away with any cosmic teleology by which the universe as a whole would be seen as ordered to some end."[11] On the Darwinian understanding, then, nature, in the cosmic or transcendent sense, did not intend and does not endorse human nature as we now possess it, because nature in the cosmic sense does not intend or endorse anything. Man is not, as religious believers might contend, the crown of God's creation, that being who uniquely among living things manifests the "image and likeness" of the divine intelligence and creativity sustaining the universe. Neither is he, as a Platonist might hold, the microcosm corresponding to the macrocosm, that being uniquely able to perceive and to order his life according to the goodness and beauty of the universe. Rather, our nature is, from the standpoint of the cosmos, merely an adaptation, or rather a combination of adaptations, that proved useful with a view to survival and reproduction in particular historical circumstances.

Cosmic nature therefore provides no basis for preferring human nature to any of the wide variety of other biological natures that evolution has generated. The various Darwinian political theorists admit that human nature is not valuable in any objective sense. "As human beings," Arnhart contends, "we naturally take an 'anthropocentric' perspective, but to do this does not dictate a cosmic hierarchy of ends." The things we desire, the goods bound up with human nature, are good for us but not good absolutely.[12] Similarly, McShea contends that because our preferences have no basis except in our species-typical feelings, "there can be no species-transcendent values," and "the statement that one species is better than another can only reflect the values of the species of the evaluator." Thus there is no "objectively qualitative difference between our species and all the others."[13] In opposition to such notions that "human rationality and freedom" are "the highest forms of life" or that "our reason can be the foundation of an ethical life superior to that of other animals," Roger Masters contends that Darwinism does not allow such "exaggerated" understandings "of human nature." On the contrary, we are "no more precious than any other living form except in our own eyes."[14]

If human nature is no better than any of the diverse natures that evolution has produced through chance and necessity, then it can be no better than any of the new natures that we might imagine and create for ourselves. If there is no cosmic teleology in light of which we can judge humanity superior to other evolved forms of life, then there is no basis for judging it superior to any new forms that might appear in the future, either through random variation and natural selection or through deliberate human invention. Thus does the Darwinian denial of cosmic teleology undermine one of the most obvious and powerful grounds on which to insist on human nature's preservation in the face of its possible technological transformation.

BIOTECHNOLOGICAL DESPOTISM

The Darwinian political theorists might respond that, while their account of the origins of human nature seems to provide no compelling reason not to manipulate it through technology, their account of human nature as the very source of values at least makes such a project appear self-contradictory and therefore unworthy of choice. How, they might ask, can there be a natural human desire to overcome human nature? Here it is necessary to recur to the distinction suggested earlier between using technology to manipulate the nature of other human beings and using it to manipulate one's own nature. With regard to the former, at least, there are evident incentives, incentives that the Darwinian political theorists would recognize as natural. The desire to exploit, to use other human beings as mere tools for ones own ends, most obviously provides an incentive to use technology to alter the nature of other human beings.

Larry Arnhart's account of slavery indicates, as we have seen, that slavery is both rooted in and opposed to human nature. On the one hand, slavery arises from the natural desire of human beings to exploit others for the sake of their own well-being and that of their kin and friends. On the other hand, slavery contradicts the natural desire of the enslaved human being to avoid exploitation. Biotechnology, however, provides those who wish to exploit with a possible solution to the dilemma created by slavery's ambiguous relationship to human nature. If slavery is unnatural, and therefore wrong, because most human beings will resist it, on what grounds would one reject slavery once it becomes possible to create beings who will not resist?

Arnhart's account of slavery even suggests (though surely without intending to endorse) such a possibility. In developing his argument that slavery contradicts the natural human desire for freedom and reciprocity, Arnhart quotes Hume, who argues that if there were "a species of creatures intermingled with men, which, though rational, were possessed of such inferior strength, both of body and mind, that they were incapable of all resistance" and could never "make us feel the effects of their resentment," then we would not "lie under any restraint of justice with regard to them." In such a case there would be only "absolute command, on the one side, and servile obedience on the other."[15] Following Aristotle, Arnhart then notes that such beings would be "natural slaves," possessing "reason to some degree, but not to the extent of being able to express their resentment at being treated as purely servile beings." They would "submit to slavery without resistance," and "the rules of justice would have no application to the master-slave relationship."[16]

As the hypothetical character of Hume's argument indicates, such beings have traditionally been confined to the imagination.[17] No longer. The modern life sciences hold that such traits as intelligence are the result of natural causes, that we are increasingly aware of the precise nature of those causes, and thus that we will soon be in a position to manipulate them at will. We should, then, be able to produce through genetic engineering such beings as Hume describes, sufficiently rational to serve our interests but not sufficiently rational to express resentment or act upon it. Indeed, the promise of modern science points to an even more perfect overcoming of the natural dilemma posed by slavery. Insofar as not only intelligence, but also emotional capacities such as assertiveness and submissiveness, arise from natural causes that will soon be under our control, it will be possible to create beings intelligent enough to do any job that most humans find distasteful yet submissive enough to do it without complaint. We could create a race of beings incapable not only of manifesting and acting upon, but also of feeling, the resentments that slavery ordinary causes. Indeed, there is no reason in principle that we could not engineer a clever subhuman race that positively enjoys its subjection.

The Darwinian account seems to forbid every possible objection to such a scheme. Arnhart notes that slavery is naturally problematic not only because it offends the slave's desire to be free from exploitation, but also because it contradicts the master's natural human desire for reciprocity. We might therefore object to such biotechnology-assisted slavery on the grounds that such arrangements would offend even the masters' sense of reciprocal justice and afflict them with a guilty conscience. As we noted in chapter 4, however, it is not clear why, on the Darwinian account, we should regard such feelings as more fundamental than the desire to exploit. It is not clear why a potential master should judge that his happiness would be better served by refraining from such exploitation than by engaging in it.

More important in this context is the doubt that the master's feelings of guilt could survive his certain knowledge that his slaves have been designed such that their slavery is not contrary to, but in harmony with, their nature. That is, a master might feel some regret for the slavery from which he benefits when he knows that the slaves hate it and hate him. But why should he entertain such feelings when he knows that they will not dislike their bondage? Masters have traditionally attempted to still their feelings of guilt by dehumanizing their slaves, both in the sense of denying their humanity in thought and word and in the sense of imposing upon them dehumanizing activities. Slavery has often produced shocking cruelties intended to confirm the slaves' supposedly subhuman status. Such dehumanization is cruel because it runs counter to the slaves' own desires for freedom and dignity. Biotechnology holds out the paradoxical prospect of humane dehumanization, that is, a literal dehumanization of the slave class that need not be viewed as cruel, because it will take away their desire to be anything but slaves.

Alternatively, one might object to such slavery on the grounds that it denies to these (modified) human beings certain experiences, of freedom and responsibility, that they ought to have. Indeed, in this context one could observe that such a slavery would be, despite its congeniality to the slaves themselves, worse in some respects than traditional slavery as it has been practiced on beings not naturally suited for it. After all, people enslaved by force and contrary to their wishes at least retain their humanity and are aware of the possibility of the human experiences of freedom and self-command. In contrast, such slaves as could be created through biotechnology would not even possess the desires for such things and therefore would not aspire to them. In any case, such an objection, because it implies that the desires and experiences of our evolved humanity are somehow objectively normative, is unavailable to those who accept the Darwinian account. We are tempted to say that such beings would have been robbed of their humanity. But if no species, and therefore no traits, are higher than any others, then it is impossible to maintain that they have been "robbed" of anything of any value.

BIOTECHNOLOGY AND THE ABOLITION OF MAN

Such a subtle and unprecedented slavery would not constitute, of itself, the complete "abolition of man," to use C.S. Lewis's phrase. For in the scenario contemplated above, man would still exist as a kind of master race exploiting, humanely, this new race of contented slaves. Some human beings might, however, be tempted to go even further, to use biotechnology to recreate their own natures, pursuing experiences that are inaccessible, or accessible only with difficulty, on the basis of our evolved humanity.

The arguments of some proponents of the new Darwinian political science seem to deny the possibility of a desire to transform one's own nature. Arnhart, for example, argues that, even in the absence of a cosmic teleology that formed nature with a view to the flourishing of beings such as ourselves, we can still be content insofar as we have evolved to live in the natural world in which we find ourselves. Thus he concludes *Darwinian Natural Right* by contending that although "the natural world was not made for us, we were made for it, because we are adapted to live in it. We have not been thrown into nature from someplace far away. We come from nature. It is our home."[18] In response to those who have suggested that "we ought to have no feelings, or that we ought to have different and better feelings," Robert McShea responds that such a notion is incoherent. "There is no basis on which we can criticize our total feeling pattern," he contends, for criticism involves a value judgment, and value judgments can have their basis only in feelings." Indeed, because our "genetic feeling pattern is our essence as a species," to wish for its alteration is to wish for our own demise.[19]

But one can as easily contend that Darwinism provides good grounds to suspect that many human beings would be dissatisfied with their humanity. Given the irrational, undirected process by which it emerged, there is little reason to hope that human nature will possess any fundamental coherence. On the Darwinian account, while all human characteristics would have evolved in the service of reproductive fitness, the behaviors and desires that serve that single end vary widely with the tremendous variety of circumstances with which living beings have been confronted in evolutionary history. One would expect, then, that human nature would consist of a jumble of contradictory drives and propensities.[20]

Such an understanding of human nature is often implicit in the arguments of the Darwinian political theorists, who present a diversity of powerful and conflicting human emotions, and who offer no persuasive grounds on which to think that some are more fundamental than others. This view is most explicit, however, in McShea's *Morality and Human Nature.* Each of the traits of an organism, he notes, evolves to some extent "in accommodation with the whole organism," but also "independently," as a "response to a particular environmental challenge," and also to some extent in "competition" with other traits for the scarce resources possessed by the organism as a whole. This is true, he continues, of an organism's feelings, and we find the feelings of a species to be "a collection of independent motivations," all of which have demonstrated their usefulness with a view to a common end, survival and reproduction, but all of which are also experienced by the organism as an "ultimate and autonomous motivations."[21] Therefore, "human feelings are necessarily in conflict," and it is impossible to satisfy them all in any of us.[22] Both McShea and Arnhart appear to suggest some measure of coherence in human nature when they contend that some feelings are deeper, more enduring, or more powerful than others. This is no doubt true, but it does not render human nature coherent. For it is very likely that there is persistent conflict even among the more powerful passions of our nature. Certain moral passions, like the desire for justice, may conflict with non-moral ones that seek to satisfy themselves using immoral means, like the desire for status—to say nothing of the conflict that exists between moral passions and ones that seem to point to immorality, say, between the desire for reciprocal cooperation and the desire to exploit.

On the Darwinian understanding, then, human experience is characterized by the incessant demand for satisfaction of a complex set of powerful and independent feelings, none of which can be gratified without mortifying some of the others. In the absence of any grounds other than their comparative strength on which to prefer some feelings to others, there is good reason to think that life with such a contradictory nature would be hard to bear. Our aim is the satisfaction of feelings, yet the satisfaction of every desire is tainted by the pain of denying some other. Moreover, because of our unusually developed intellectual capacities, we are all too keenly aware of these difficulties. Once again, this point, implicit in the Darwinian account of human nature, is made most explicit by McShea. Because of our intellectual capacities, McShea notes, we can experience not only what is happening around us but also our own thoughts, our imaginings of what has been and what might be. Moreover, we react to both reality and our imagination with passionate feelings, and as a result human desire is infinite.[23] While our imagination sets us apart from every other species, we suffer on account of it. Only humans, McShea writes, "experience serious functional psychic disorders in our ordinary social life. These disorders, which arise from our feeling conflicts, are the principal cause of human unhappiness, outranking disease, death, and physical deprivation."[24]

McShea offers ethical theory as a solution to the human problem. By careful self-examination we can discern our most important
and enduring feelings, and by self-control we can pursue them in
preference to others, thus achieving, in the end, more satisfaction
than if we had succumbed to insistent but passing desires. McShea's
account also indicates, however, the necessary imperfection of such
a solution. If human desire is infinite but human life finite, then
even the most successful ethical life will involve considerable dissatisfaction. Moreover, even the limited happiness to which we can
aspire is very difficult to achieve. Because we cannot eliminate our
intellectual capacities and the infinite desire that accompanies them,
McShea contends, the easy contentedness that "other animals achieve
effortlessly, we must strive for along the hard route of delay, reflection, self-control, and the recognition of harrowing ambiguity."[25]
This route is sufficiently hard, McShea observes, that many humans
choose not to take it, opting instead for the (ultimately even less
satisfying) pursuit of immediate pleasures or submitting themselves
to some social authority that will determine their actions for them.[26]
Indeed, McShea suggests that all human beings on some level and at
some times yearn to relinquish their humanity, and thus he implies
the unsatisfactory nature of the ethical and good life even for those
who are able to achieve it.[27]

In this light, the Darwinian account seems not merely to remove
all grounds for objecting to technologically manipulating human
nature, but in fact openly to invite it. If we know that our humanity is not the result of some cosmic design but only of evolutionary
happenstance, if we experience our humanity as a burden, and if it
is now, or soon will be, in our power to change our humanity, then
why should we not change it? If there are aspects of our nature that
we find troublesome, why not excise them through biomedical manipulation as soon as possible?

In the absence of some cosmic teleology that can account for the ultimate goodness of our hard condition, Darwinism can only offer prudential arguments against such modification. The genetic constitution of a species, the Darwinian can argue, is like an ecosystem: one cannot alter one element without of necessity altering others, often in unpredictable ways.[28] For example, because we evolved to cooperate within groups in order to compete with other groups, our capacity for social bonding may be genetically bound up with our propensity for aggression, such that it would be impossible to excise the latter without impairing or destroying the former.[29] Such arguments would seem to be of limited utility, insofar as they have been advanced in opposition to every successful new technology ever devised. More important, without reference to some cosmic teleology—in light of which one might judge human nature, and hence certain human traits—as objectively valuable, there is no basis on which to say that someone has incurred a loss who has suffered the unintentional impairment of one emotional capacity as the result of the deliberate elimination of another. In any case, such arguments are not principled objections, and they would lose their salience if our scientific mastery were to prove equal to the task discerning all possible consequences of genetic engineering and of excising some traits while preserving related ones.

Moreover, it may be that the biomedical transformation of our humanity will not require such radical methods as genetic engineering. As Francis Fukuyama points out, the issues that arise from the possibility of manipulating human nature have "already been joined with the current generation of psychotropic drugs, and will be put into much sharper relief with those shortly to come."[30] Perhaps, then, we will discover the possibility of an untroubled happiness on the basis of neuropharmacology. If we find that it takes hard work to satisfy our feelings through an ethical life, and if we agree with the

Darwinians that a happy and good life is simply one with the maximum possible satisfaction of feelings, why not seek ways to satisfy them artificially and to make this the basis of our happiness? If we posit, for example, that the sense of satisfaction that comes from some difficult accomplishment, or from winning the esteem of one's fellow citizens, is merely the result of certain chemical processes in the brain, and if we learn how to stimulate such processes artificially through drugs, why not do it, as an easier way to the happiness that until now has only been available after enduring the pain of much self-denial?

One might object that anyone who avails himself of such a drug would be preferring illusion to reality. But if happiness is the satisfaction of feelings, what difference does it make if they are satisfied through one means or another? Indeed, it is not clear why one should prefer reality when, as McShea points out, it is so difficult to achieve happiness under the "real" conditions with which we have had to contend throughout most of our evolutionary history. The preference for reality is further undermined by modern evolutionary science's suggestion that human consciousness is itself illusory. According to Roger Masters, contemporary "biology teaches us that," from "the perspective of the cosmos," the "primary reality may not be the physical or existential realm of which we are conscious, but rather the gene pool that we inherit and transmit to our descendants."[31] Finally, since Darwinism denies any notion of a cosmic teleology that somehow intended to produce human consciousness, one wonders how we could rationally prefer, or even regard as more "real," a satisfied state of consciousness that arises from active engagement with the surrounding environment in comparison to a satisfied state of consciousness that arises from the artificial chemical stimulation of the brain.

One might object to a drug-induced happiness on the grounds that to seek happiness apart from personal effort is effectively to negate free will. But modern evolutionary science contends that our subjective consciousness of free will is just as illusory as consciousness generally. The orderliness of nature "is determined," McShea contends, "and although the determinism we insist on finding in nature becomes ever more complex and obscure as we go from atoms to cells and from apes to humans it nevertheless remains true that, when we say that we are a part of nature, we mean that we are committed to finding ourselves exhibiting lawlike characteristics."[32] On this view, human behavior may be unpredictable, because it is the result of a multiplicity of difficult to assess causes, but it is determined nevertheless. Any human being simply "does what he must, given his feelings and perceptions," which are themselves the result of a deterministic chain of causality and hence not the responsibility of the actor.[33] McShea concludes that we "cannot confer a metaphysical moral responsibility on others," but that we are nonetheless justified in blaming and praising them provisionally, because we have evolved a desire to do so and because praise and blame can encourage behavior we like and discourage its opposite.[34] Roger Masters similarly contends that the more we learn from "ecology, neurobiology, and molecular genetics, the presumption that human behavior is uncaused or controlled by free will becomes less and less tenable."[35]

The Darwinian political theorists make much, as we have seen, of Darwinism's refutation of Hobbes. They do so with justice, for Darwinism certainly does deny, and with persuasive evidence, Hobbes's notion that humans are by nature asocial and that morality is a mere social construct. Hobbes also held more generally, however, that the universe is fundamentally chaotic, that happiness is the satisfaction of desire, and that human desires are fundamentally contradictory —notions that open the door to and openly invite the overcoming of human nature through human ingenuity, through a kind of technology. Darwinism is decidedly Hobbesian in this, perhaps more fundamental, sense, for it shares Hobbes's understanding of the whole and of man, and it points to the same aspiration for self-transformation.[36]

While Hobbes had to rely on the crude technology of an all-powerful state, modern science offers the possibility of a more effective solution, one that does not merely suppress through intimidation the display of certain passions, but that prevents them from being felt at all by altering the genome or manipulating the brain.

FUKUYAMA'S CRITIQUE OF BIOTECHNOLOGY

Despite Darwinism's relative congeniality to the conquest of human nature, the new Darwinian political theorists are not notably enthusiastic about such a project. Some, indeed, have expressed reservations about it. In *The Nature of Politics*, Roger Masters gives voice to fears that, in the absence of clear moral standards, we might use our emerging power over nature to create genetic monsters.[37] This is not a question to which he gives much attention, however. James Q. Wilson has argued for some restrictions on human cloning, though such limits as he envisions are quite modest, and his argument for them does not indicate a principled opposition to such uses of biotechnology per se.[38]

Among the Darwinian political theorists discussed in this book, Francis Fukuyama has devoted the most attention to the possible biotechnological conquest of human nature. His most recent book, *Our Posthuman Future*, outlines the various threats to our humanity posed by neuropharmacology and biotechnology. He argues that democratic communities must draw lines confining the applications of biotechnology to uses that are not harmful. He contends that it is possible, contrary to those who claim that technology cannot be controlled, to develop a regulatory framework that allows some applications, controls others, and bans still others. To this extent his work is welcome, since it might give courage to those who, while concerned about the possibilities biotechnology presents, have resigned themselves to the impossibility of subjecting it to political control. The difficulty with *Our Posthuman Future* is that Fukuyama's arguments against the conquest of human nature are not compatible with the Darwinian understanding he adopts in *The Great Disruption* and reaffirms in this most recent work. Thus his arguments implicitly confirm, contrary to his intention, the continuing need for religion as a source of moral wisdom.

Fukuyama identifies utilitarian, religious and philosophical arguments by which one might object to the conquest of human nature. The utilitarian objections center on the various unforeseen and undesirable side-effects that might attend efforts to alter human nature. While such concerns are worthy of consideration, and have the advantage of being readily intelligible to the public, Fukuyama suggests that they are limited because they fail to take into account the threat biotechnology presents to our humanity itself, the dangers that it poses "to the soul rather than to the body."[39] Religion, on the other hand, is preoccupied above all with the well-being of the soul, and therefore it is not surprising that it "provides the clearest grounds for objecting to the genetic engineering of human beings." In the religious "tradition shared by Jews, Christians, and Muslims," Fukuyama points out, "man is created in God's image," possessing unique capacities for "moral choice, free will, and faith" that gives him "a higher moral status than the rest of animal creation."[40] Thus religious believers tend to reject biotechnology on the grounds that it represents an impermissible and hubristic tampering with the loftiest part of God's creation. While religion provides strong reasons to oppose the transformation of human nature, it does not necessarily provide reasons that will move the "many" citizens "who do not accept religion's starting premises."[41] An objection to biotechnology that is both principled and publicly persuasive would seem to be beyond our reach, then, were it not for the possibility of a philosophic argument that presents human nature itself as the source of human dignity and of our moral knowledge.

Fukuyama points out that while Pope John Paul II appears to have accepted the notion that human beings evolved, he nonetheless contends that humanity could only emerge from the process of evolution by means of an "ontological leap" that distinguishes man radically from the other animals. The notion of such a leap, moreover, is in the pope's view essential to the belief that humanity is possessed of a special dignity: "theories of evolution which . . . consider the mind as emerging from the forces of living nature, or as a mere epiphenomenon of this matter, are incompatible with the truth about man" and unable to "ground the dignity of the person."[42] Fukuyama contends that the pope is correct that an "important qualitative, if not ontological, leap" was necessary for man to emerge, that man is unique within nature, that such propositions are "fully compatible with modern natural science." Therefore, one need not accept the pope's belief that the "human soul" is "directly created by God" in order to embrace them.[43]

While much modern science is reductionistic, explaining natural phenomena as mere manifestations of the behavior of their constituent elements, Fukuyama contends that such reductionism cannot explain the behavior of many things in nature. Natural science now understands that the interaction of parts in complex systems—such as the behavior of groups of sociable animals acting in common—often gives rise to a whole that is qualitatively different from, as well as unpredictable on the basis of, the behavior of the parts. Man enjoys a "place at the uppermost level" of the "hierarchy" of complexity in nature, and therefore his nature is especially immune to a reductionistic treatment. Distinctly human capacities—for political interaction, for rational speech, for self-consciousness itself—cannot, on this view, be completely understood as mere products of man's biological and chemical components, nor as completely reducible merely to more complex manifestations of their non-human analogues. Even our emotions, Fukuyama contends, are not completely comprehensible on reductionistic grounds, which can explain their functions but not the "particular subjective form" that they take.[44]

Human beings, Fukuyama concludes, are set apart from the rest of nature by our being "complex wholes rather than the sum of simple parts," and this confers upon us a "dignity and moral status higher than that of the other living creatures."[45] This understanding of the radical distinctiveness of humanity is the key to Fukuyama's argument against the possible transformative applications of biotechnology. We must seek to protect our unique and uniquely valuable complexity from a biomedical science that threatens to simplify our nature in the utilitarian attempt to gratify particular desires, Fukuyama argues. For modern biomedicine often tends to "reduce" the "complex diversity" of our "natural ends and purposes" to "simple categories like pain and pleasure, or autonomy." "There is in particular a constant predisposition to allow the relief of pain and suffering to automatically trump other human purposes and objectives," such as the exercise of "moral choice, reason, and a broad emotional gamut." Thus biotechnology poses a "constant trade-off": "we can cure this disease, or prolong this person's life, or make this child more tractable," but only "at the expense of some ineffable human quality like genius, or ambition, or sheer diversity."[46]

Fukuyama's argument for the preservation of our inherited nature is undermined, however, by his apparent embrace of the Darwinian denial of cosmic teleology. As noted earlier, in *The Great Disruption* he contends that all natural order is the result of the blind and irrational operation of natural selection.[47] He seems to reaffirm this view in *Our Posthuman Future* when he notes that any species essence is merely "an accidental byproduct of a random evolutionary process" and that the human brain "arrived at its present state through" a process of "random variation" and "natural selection."[48] Without cosmic teleology, in the absence of any species-transcendent account of ends or goods, there are no grounds on which to attribute to a certain species a value objectively superior to that of other species, or indeed on which to attribute to any species any objective value at all. Fukuyama's argument may demonstrate that human nature is, by reason of its complexity, qualitatively distinct from all other animal natures, but it cannot show that the substance of our complex human nature is superior to that of the other various animal natures. Put bluntly, Fukuyama's contention that we have a special dignity because we have a unique complexity is, on his own theoretical principles, a non sequitur.[49] On the Darwinian account that he follows, human-ity must be judged as different from but no better than the rest of living nature, and therefore there can be no compelling reason to preserve human nature, especially when confronted with powerful desires to modify it.

Fukuyama's rejection of cosmic teleology undermines his objec-tions to particular applications of biotechnology. For example, he condemns reproductive cloning as a practice that "should be banned outright" because it is "a highly unnatural form of reproduction that will establish equally unnatural relationships between parents and children."[50] But such an argument depends on the assumption that our evolved nature, and the relationships to which is has historically given rise, is somehow objectively normative, an assumption that Fukuyama's Darwinian principles forbids. He also opposes cloning on the tactical grounds that it opens the door to the genetic engineer-ing of "designer babies," which also ought to be forbidden. Genetic manipulation of human beings, he argues, should be allowed only for therapeutic purposes, to correct defects and eliminate disease, and not for "enhancement," to improve upon normal human capacities. Such a distinction is possible on the view that human health is normative for medical science, the aim of which should be only to cure the sick and heal the injured. Health, Fukuyama suggests, is a natural stan-dard, the "natural functioning" of "the whole organism that has been determined by the requirements of the species' evolutionary history," and not merely "an arbitrary social construction.[51] Again, however, it is unclear why we should be obliged to so limit the aims of biotechnology when the natural functioning of our species is only the result of evolutionary accident and also often unsatisfactory to us.

Fukuyama advances two other general arguments against the biotechnological enhancement of human nature, both of which are undercut not only by his by the Darwinian denial of cosmic teleology, but also by the Darwinian account of human nature. First, he warns that the pursuit of such enhancement may destroy the grounds for our belief in the fundamentally equal dignity and rights of all human beings. It is possible that genetic enhancement will be too expensive to be widely available, and that as a result genetic excellence—in the form of superiority in traits like "beauty," "intelligence," "diligence," and "competitiveness"—will more and more reside in the social elite, creating a "genetic overclass." The members of such an elite "will look, think, act, and perhaps even feel differently" from lesser mortals, and thus they may look upon themselves as a different kind of creature, a natural aristocracy with superior claims to rights and deference.[52] After all, Fukuyama argues, our commitment to the fundamentally equal dignity of all human beings has always been based on the assumption that all human beings have fundamentally the same nature.

Because the loss of cosmic teleology obliterates any objective grounds on which one could value such equality of essence and of dignity, Fukuyama must appeal to what he takes to be a powerful human desire to preserve the now common respect for equal human rights. With the same historicist confidence he displays in *The Great Disruption*, Fukuyama contends that, although humans have for most of history recognized the dignity of only selected other human beings, and although the religion that introduced the notion of a universal human dignity is now on the decline, respect for the equal dignity and rights of all human beings is now firmly rooted in our culture.[53]

As we noted in chapter 4, however, it is extremely doubtful, on the Darwinian understanding, that an inclination to respect the dignity of human beings as such would be very powerful, insofar as our moral nature was evolved under conditions in which morality within groups was pursued precisely with a view to gaining an advantage in an amoral competition among groups. It is true, as Fukuyama points out, that the progress of our science has revealed our common human nature and thus shown that "the grounds on which certain groups were historically denied their share of human dignity" were merely prejudices or "based on cultural and environmental conditions that could be changed."[54] Despite our intellectual embrace of the equal claims of all humans, however, on the evolutionary understanding one may reasonably expect that the desire to respect the dignity and rights of others, especially of others who are not members of the social group with which one identifies, will be considerably weaker than our desire to advance our own interests and those of our children. And biotechnology's potential to enhance such traits as intelligence and competitiveness appeals to one of our most powerful evolved desires—for high social status and the resources that accompany it. On a moral theory that seeks to derive norms from our evolved feelings, genetic enhancement must appear, to any given individual, as a greater good than the equality of human dignity.

Fukuyama also objects to biotechnology because, as a result of the utilitarianism that typically guides its application, it will probably be used single-mindedly to pursue the greatest possible minimization of human suffering. While "no one can make a brief in favor of pain and suffering," he argues, "what we consider to be the highest and most admirable human qualities . . . are often related to the way that we react to, confront, overcome, and frequently succumb to pain, suffering, and death. In the absence of these human evils there would be no sympathy, courage, heroism, solidarity, or strength of character." Fukuyama concludes his book by warning that our posthuman future "could be the kind of soft tyranny envisioned in *Brave New World*, in which everyone is healthy and happy but has forgotten the meaning of hope, fear, or struggle."[55]

Once again, the grounds for such objections are destroyed by Darwinism's understanding of both the cosmos and of human nature. For if the moral excellences Fukuyama praises are merely (up to now) necessary reactions to the evils imposed on us by a mindless evolutionary process, how can they have any inherent or objective value? Moreover, as we discussed in chapter 3, the account of human nature offered by the Darwinian political theorists places almost all of its emphasis on morality as a mere decent sociability and gives us little reason to expect that a desire to practice the more difficult virtues would be a powerful part of our natural emotional constitution. This moral minimalism is evident in *The Great Disruption*, and it appears again in *Our Posthuman Future*, particularly in Fukuyama's praise for liberal democracy. Liberal democracy, Fukuyama suggests, is the political regime most in harmony with human nature. He contends, for example, that an "important reason" for the "worldwide convergence on liberal democracy" is "the tenacity of human nature." Elsewhere he holds that the "findings" of "modern science" indicate that "contemporary capitalist liberal democratic institutions have been successful because they are grounded in assumptions about human nature that are far more realistic than those of their competitors."[56]

Yet as Tocqueville suggests, and as Fukuyama surely knows, liberal democracies are also extremely preoccupied with the use of technology to minimize suffering and create conditions of ease and comfort. Moreover, as Fukuyama observes in *The Great Disruption*, liberal democracies "buy political order at the price of moral consensus" —that is, they are founded on a kind of agnosticism about the very virtues that Fukuyama, in his critique of biotechnology, treats as of acknowledged public significance.[57] Indeed, in *The Great Disruption* Fukuyama attributes liberal democracy's success precisely to the fact that it recognizes that "[i]ndividual self-interest is a lower but more stable ground than virtue on which to base society."[58]

Fukuyama's account of human nature, then, suggests that our desire for ease, which requires the continual removal of our limitations, is far more powerful than our longing for moral excellence or nobility of spirit, which is displayed in facing our limitations. Fukuyama, like the other Darwinian political theorists, seeks to ground our understanding of the good on our evolved desires. Thus his argument suggests, contrary to his intention, that the progressive relief of suffering that biotechnology offers is better than the moral nobility the exercise of which it continually impairs.

CONCLUSION

Darwinism invites the technological transformation of human nature by its denial of cosmic teleology and its denial that human nature possesses any coherence in light of which we could forge any fully satisfactory happiness in this life. Of the two denials, the former is by far the more significant. That human nature is in some sense chaotic, that we are constantly moved by contradictory desires that drive us toward incompatible goods, is not merely a tenet of Darwinism but also the understanding of many religious traditions and, indeed, an elementary fact of human life that we all recognize. It is not so much in pointing out that human life is beset with difficulties that Darwinism invites the "abolition of man," but in its assumption that the difficulties, and therefore the virtues that we display in contending with them, are ultimately of no objective worth. They are merely the burdensome necessities of living with a nature that is itself merely the product of a combination of chance and necessity. Once it becomes possible to tamper with such a nature, such a notion of the cosmos can provide no compelling reason for not doing so.

One obvious source of a cosmic teleology, however, is religious belief, on the basis of which one may view our struggles with our problematic humanity not only as burdensome but also as objectively meaningful—and perhaps, as a way of glorifying God, even bound up with the ultimate meaning of the universe. Despite their similarities, there remains a decisive theoretical—and, in the face of the possible conquest of human nature, practical—difference between Francis Fukuyama and John Paul ii. While both recognize the uniqueness of man, only the latter can see that uniqueness, with all of its miseries and glories, as part of the aim of a cosmic design that is good, not as a chance outcome of an evolutionary process that is ultimately unintelligible. Between the two, only the latter can make a convincing case for the preservation of human nature. Here again, then, on the threshold of the greatest moral dangers, religion reveals its continuing public importance to modern democracy.

8

Beyond Darwinism

T HE INSIGHTS OF EVOLUTIONARY BIOLOGY provide a certain
support for conservatism's concern with justifying the morality necessary for a decent and freely cooperative society.
Darwinism offers some support for morality generally, and thus offers
a kind of response to moral relativism, by showing that our sociable
and moral impulses—for bonding, esteem, sympathy, reciprocity—are
rooted in our biological nature and not merely the products of social
learning. A Darwinian conservatism can point out, with the authority of science, that when a person acts in an immoral or anti-social
manner, he will not simply be throwing off the artificial constraints
that society has imposed in contradiction to his nature (unless he is
a psychopath), but that he will be acting contrary to an important
part of his nature, and therefore that he will likely pay a serious and
natural emotional cost in guilt and shame. Conversely, when someone
behaves in a moral and sociable way he will be satisfying powerful natural desires. Darwinism appears to offer a kind of answer to
Glaucon and Adeimantus' challenge to Socrates in Plato's *Republic*:

that Socrates show that justice is good for the soul and not merely for its desirable consequences.[1]

In addition, the Darwinian account of human nature provides support to a number of conventionally conservative positions. Insofar as it shows that the inclination to give parental care to one's own biological offspring is rooted in our nature, Darwinism suggests that what has often been called (even by conservatives) "the traditional family" might more properly be termed "the natural family." It is likely that conservatives are correct in their contention that the nurturing of children will be done, all other things being equal, more effectively, because with more natural interest, by the biological parents than by someone else. Moreover, Darwinism's contention that psychological differences between the sexes—that women tend to be more attached to children and more inclined to nurturing and that men tend to be less attached to children and more competitive —are grounded in our biological nature lends support to the conservative preference for traditional sex roles. Such an insight does not justify imposing such roles on men and women through legislation (which, in any case, is favored by no conservative of whom I am aware), because the differences exist in the aggregate, and therefore the disposition of any particular man or woman may not conform to such roles. It does, however, tend to justify the actions of those who choose voluntarily to adopt those roles. A woman who elects to treat mothering as more important than a career would be properly understood not as slavishly living out a false consciousness imposed by a patriarchal society, but instead as understandably acting according to her own powerful and natural desires.

Darwinism also suggests, as John McGinnis notes (offering, in the pages of *National Review*, perhaps the best short summary of Darwinism's support for conservatism), the natural status of private property, market exchange, and the inequality of wealth. A number of the elements of our evolved nature point in this direction. As

McGinnis contends, Darwinism indicates that self-interest will be deeply rooted in human nature. Human beings, after all, will want resources for themselves and their offspring so that they can be used to advance their natural interest in reproductive fitness. Moreover, humans have, as Arnhart and the other Darwinian political scientists also point out, evolved a desire to advance their interests through freely undertaken reciprocal exchange, such as occurs in markets. In addition, McGinnis notes that modern biology holds that "individuals have inherently unequal abilities and that these inequalities are likely to be greatest in the personality traits, such as intelligence and ambition, that are related to acquiring property."[2] The Darwinian understanding of human nature, then, points to the conservative conclusion that the inequalities of wealth resulting from free, market-based exchange are simply the consequence of nature taking its course, and not the unjust fruits of an artificial economic system that fosters exploitation. Thus Darwinism confirms the conservative wisdom of American Founders like James Madison, who sought a system of institutions that would protect the natural inequalities of property arising from our naturally unequal capacities.

Finally, in its general affirmation of human nature as a universal and permanent reality, Darwinism seems to suggest good grounds for limiting the power and ambition of the state. Conservatives contend that much harm has been done, especially in the last century, as a consequence of state-sponsored social engineering that sought to turn human beings into something other that what they are. Such aspirations seemed reasonable to the social visionaries who undertook them because they believed that what human beings are is the result not of nature, but of social circumstances. Change social conditions, it was thought, and you can change man. While acknowledging the role of environmental factors in shaping human behavior, Darwinism nevertheless shows that human behavior is to a considerable extent the result of biological propensities that cannot be eradicated by altering the social environment. Darwinism suggests that human nature is not infinitely malleable. Attempts to mold human nature much beyond its ordinary manifestations will run counter to powerful innate desires and therefore generate great misery.

As we have seen, the Darwinian account of human nature is of limited usefulness from the standpoint of Tocqueville's conservatism, which seeks to conserve the conditions of human dignity and freedom within democracy by resisting the democratic inclinations toward decent materialism and tyranny of the majority. The morality that Darwinism seems able to justify as rooted in human nature appears to be a morality of mere decent, sociable cooperativeness. Such a morality is itself compatible with the democratic hedonism that Tocqueville deplores, and so it seems that Darwinism offers nothing on the basis of which one could support Tocqueville's desire to encourage men to aspire to a greatness that transcends the petty aims of ordinary democratic life. Moreover, the Darwinian account of human nature and the conditions under which it evolved casts grave doubt upon the strength of any natural human desire to respect the rights of other human beings, as human beings, when one's own group's interests could be advanced by violating them. Thus Darwinism provides little support for Tocqueville's efforts to encourage such a morality as would moderate the tyranny of the majority.

Moreover, the critique offered in the preceding chapters also indicates that Darwinism is of limited value even with a view to the ideological uses to which some contemporary conservatives want to put it. Consider the example of the family. While it is true that Darwinism shows the natural basis of the family, it also shows that the family—at least as conservatives desire and as society needs it—is not fully natural but depends on some measure of social constraint. On the one hand, the adult's natural desire to provide parental care, the child's natural need for parental care, and the requirements of social stability all lend support to social policies that foster the stability of the familial bond. Such policies help fulfill important natural desires and interests. On the other hand, as we saw in chapter 5, to the extent that males are by nature much less attached to their mates or their children—to the extent that they have powerful promiscuous desires that seem to drive them away from the family—such policies can be viewed as constraints on the nature of men. It may be true, as McGinnis argues, that "the new biological learning provides direct support" for conservative criticism of public policies like no-fault divorce and welfare. These policies make it extremely easy for men to abandon the family in search of new sexual opportunities, and conservatives generally favor policy changes like welfare reform and covenant marriage that tie men to the family. Nevertheless, it is equally true on Darwinian grounds that fathers are naturally "more likely than mothers to resent and avoid obligations that may deprive them of other mating opportunities." Therefore, on Darwinian grounds alone one must conclude that when men try to resist or escape the consequences of family-friendly policies they will only be doing what comes naturally.[3] The Darwinian account cannot finally tell us why we should sacrifice the natural desires of men to those of women and children.

Alternatively, consider the Darwinian conservative defense of private property, market exchange, and inequality. It is true, as McGinnis argues, that human beings possess natural inequalities in ability that, along with their natural desire to engage in reciprocal exchange with a view to their own self-interest, give rise to private property possessed in varying amounts. It is also true, however, as we saw in chapter 4, that human beings have a natural desire to advance their interests through non-reciprocal exploitation, as well as a relatively weak natural interest in the well-being of those beyond their kin and social group. When humans try to expropriate, through force or fraud, the freely acquired property of others, they may be seen as acting in accordance with their natural propensities. Again,

McGinnis recognizes this aspect of our nature. "Biologists now recognize," he writes, "that one of the strongest primate tendencies" is "the drive to form coalitions that can seize the social surplus for" the benefit of coalition members.[4] Of course, one can, with McGinnis, deplore such tendencies as "not worth celebrating," and contend that our knowledge of them points to the wisdom of institutions—such as limited government—that prevent the social disruptions and diminished general prosperity to which such behaviors lead.[5] But if these propensities are among the strongest in our nature, they are presumably as authoritative for us as our inclination toward reciprocal exchange. From a purely Darwinian standpoint, those who try to use the power of the state to expropriate the superior wealth of others behave no less naturally and reasonably than those who acquired their superior wealth through voluntary market exchange. In sum, if justice is rooted in our nature on the Darwinian view, so is theft.

As this last point suggests, even Darwinism's generalized support for morality as rooted in human nature begins to look shaky on close examination. As we noted in chapter 6, psychopaths are rare, and therefore most of us experience the normal moral sentiments of sympathy and justice. At the same time, it is very likely that most of us harbor, alongside our moral inclinations, Machiavellian inclinations, desires to cut moral corners, to advance our interests and satisfy our desires through unscrupulous and unfeeling manipulation, when we think we can get away with it. Thus Darwinism's support for morality must be regarded as ambiguous, since Darwinism takes its bearings from our evolved human nature, which it admits has both moral and immoral inclinations. The only solution to this difficulty, short of admitting some principles of moral judgment not derived from our evolved feelings, is to hold, with Arnhart and McShea, that somehow our sociable and moral tendencies are ultimately stronger, more enduring, or more deeply rooted than the others. As I suggested earlier, this seems far from obvious.

Finally, then, Darwinism cannot give unequivocal support to any political or moral position because its account of human nature points to a wide variety of different and even contradictory natural behaviors. This diversity within human nature cripples Darwinism's ability to give rise to any coherent moral or political theory because it tries to rely solely on its account of our evolved nature, our natural desires and propensities, as its source of moral and political insight. It lacks some principle—indeed in its most thoroughgoing formulations denies the possibility of any principle—beyond our feelings in light of which we can make moral judgments about the moral priority of our feelings.[6] Traditionally, such principles were sought in cosmic teleology, in the notion that the universe is ultimately coherent, that the human mind can grasp its intelligible order, and that this order includes principles of morality or justice in light of which we could judge the worthiness or nobility of our desires, irrespective of their natural strength. Darwinism rejects the notion of cosmic teleology.

This denial becomes most ominous, as we have seen, when we confront, as we apparently soon will, the possibility of the conquest of human nature. As McGinnis observes, the incoherence of human nature—and the inevitably problematic character of the human condition—to which Darwinism points can be interpreted as supporting a kind of conservative wisdom about the limitations of politics and of all human endeavor. Darwinism, he contends, holds that the human "self" is "fragile and divided." "The self, like all essential aspects of man, is an adaptation to selective pressures over millions of years and is thus jury-rigged from different mechanisms from our evolutionary past." Our various selves, like the sexual and acquisitive, "evolved for different purposes and are not fully connected." Darwinism therefore undermines utopianism, and even "utopian conservatism," the belief that there is some arrangement of things "in which all possible human goods" can "be fully and equally realized." Thus evolutionary theory, by teaching that "the different adaptations around which emotions are structured are inevitably in conflict," therefore "shines a somewhat tragic" and sobering "light on the desire for perfection in human affairs."[7] While this argument is fine as far as it goes, as we saw in chapter 7, when one combines a belief in the burdensome incoherence of human nature with the Darwinian denial of cosmic teleology one opens the door to, and actively encourages, the least conservative undertaking yet imagined: the technological transformation of human nature with a view to eliminating our apparently purposeless suffering. While undermining the leftist's desire for social engineering, the evolutionary account invites the even more radical enterprise of wholesale biological engineering.

The key to the failure of Darwinian morality is its lack of a cosmic teleology. In the absence of some principles of goodness that transcend our nature as Darwinism presents it, there can be no intelligible reason to prefer our noblest but weakest desires to our strongest and most commonplace ones, to prefer our strong and decent desires to our equally strong but unscrupulous ones, or to prefer our natural but trying humanity to an artificial and easy post-humanity. That is, without cosmic teleology we cannot vindicate Tocqueville's concern with conserving the possibility of human greatness and a principled respect for rights, we cannot conserve even the ordinary decency and social order with which the Darwinian conservatives are preoccupied, and we cannot conserve our own humanity in the face of attempts to "enhance" it out of existence. One obvious source of cosmic teleology is religion, which can render intelligible the notion that some of our impulses, though weaker than others, are nobler and more obligatory, and that our fractured humanity is, though certainly a source of grief, also an occasion for glory. In light of the failure of Darwinian morality, religion retains, in accord with Tocqueville's understanding, its political and moral importance—at least if one is interested in conserving our capacity for nobility, for justice, and even our humanity itself.

One might object to this turn to religion as an introduction of irrationality into morality and politics. This objection is implicit in Larry Arnhart's criticism of Andrew Ferguson for his repudiation, in the *Weekly Standard*, of Fukuyama's presentation, in *The Great Disruption*, of morality as rooted in our evolved human nature. Ferguson, Arnhart notes, attacks scientific materialism as a kind of myth and advises conservatives to rely "on 'the older myths' of free will and natural law as the intellectual foundation for their moral and political thought." Arnhart argues that "Conservatives such as Ferguson, who reject a theoretical foundation in human nature, must ultimately appeal to 'myth' as their final ground of judgment, which follows the lead of such conservative thinkers as Richard Weaver who spoke of the 'metaphysical dream' of transcendent order as a poetic creation necessary for any culture. The danger here is that conservatism begins to look like a Burkean Nietzscheanism, in which the moral order of society requires mythic traditions as noble lies that hide the ugly truth of nihilism."[8] Arnhart thus suggests that the introduction of cosmic teleology as a basis for morality will necessarily be non-rational, whether or not it is expressly rooted in revelation.

This charge presents a serious practical and theoretical challenge to any transcendent, as opposed to evolutionary, conservatism. From the standpoint of a practical interest in public morality, one must be concerned with what kind of arguments will likely prove persuasive to most people. As even Tocqueville admits, democratic men are, because of their instinctive rationalism, especially unlikely to be moved by appeals to myth-based religious authority. From the standpoint of the theoretical concern with the truth, one might ask why we should accept religion's revelation of cosmic teleology even if is necessary to sustain society's commitment to morality. After all, that a belief serves some practical purpose, even a good purpose, does not make it true.

In response to Arnhart's objection we may contend that this turn to revelation need not be seen as a simple rejection of science, an embrace of myth, or a surrender to irrationality. In the first place, one may accept revelation—at least the Christian revelation—as true, and hence take one's moral bearings from it, without rejecting the insights into biological nature generally, and into human nature specifically, that are offered by contemporary science. It is evident that all living things die, that successful reproduction is not easy, that inherited traits vary, and therefore that there will be selective pressures on species such that their traits will tend to be modified with a view to reproductive fitness. There is nothing in revelation to contradict these obvious facts, and therefore religious believers can accept that the physical and even the emotional and moral constitution of human beings has been shaped by natural selection. Moreover, in accepting, and in trying to adhere to, a religious account of morality one need not ignore modern biology's insights into our innate proclivities. Even if we do not take the Darwinian account of human nature as the source of our moral aspirations, we may prudently take it into account as limiting what we can reasonably hope to achieve.[9]

Arnhart might contend that, even in light of such qualifications, the turn to a religiously-informed cosmic teleology is fundamentally the embrace of an irrational morality: for those who make such a turn will derive their moral aspirations not from a rational account of human nature but from a revelation that is not in principle knowable by all men. Thus they step beyond the realm of reason and into that of myth. In response, I would contend that such a turn is not merely an abandonment of reason, because it is more reasonable to understand ourselves in light of both the Darwinian account of our nature *and* the religious account of our nature and the cosmos, than it is to understand ourselves in light of the former in the absence of the latter. For the Darwinian account, despite its legitimate insights into some elements of our nature, cannot fully explain what we are, cannot completely account for the full range of our moral aspirations —especially that for moral nobility, which goes beyond anything that might contribute to our reproductive fitness but which we nonetheless experience as an essential element of our being.

In a critique of a number of recent books on evolutionary psychology, Andrew Ferguson complains that Darwinism cannot explain Mother Theresa, whose commitment to morality extended far beyond an adaptive decency and cooperativeness and whose capacity for sympathy seemed to extend to any human being.[10] Of course, as was suggested in chapter 3, Darwinism *can* offer a kind of explanation of her: as an evolutionary fluke, a kind of reverse-psychopath with a remarkable hypertrophy of the moral impulses. More precisely, then, what Darwinism cannot explain is our genuine admiration for Mother Theresa, our experience of her as not just a behavioral curiosity —though perhaps a useful and pleasant one—but as a moral exemplar, a living sign of what we could be and should strive to be.

The Darwinians might object that this argument simply begs the question by assuming the natural value and intelligibility of the moral excellence endorsed by religion: it may be that whatever admiration people feel for Mother Theresa is the result of cultural prejudice, of indoctrination in religious morality, and not of a natural human inclination toward the noble that transcends our evolved desires. Perhaps, however, all human beings would concede, on deep reflection, the poverty of an account of man incapable of affirming that moral heroism is by nature not only rare but noble, not only useful but beautiful, not only good because it serves our needs but also because it stimulates our emulation. Moreover, as I have argued throughout this book, the Darwinian political theorists themselves cling tenaciously to moral aspirations—for nobility and universal justice—that cannot be adequately defended on the basis of their evolutionary naturalism. Indeed, as Arnhart points out, Darwin himself, even as he contended that everything about man could be explained in light of evolutionary causes, harbored a longing for universal humanitarianism incompatible with his own scientific account of the development of the moral sense. Thus the thought of the Darwinians themselves provides evidence of Darwinism's incapacity fully to explain man and morality, and its consequent inability fully to explain nature as it is. This insufficiency invites us to consider anew an understanding that can embrace Darwinism's insights while supplying its deficiencies.

Notes

Preface

1   Pinker 2002, 1-2.
2   Quoted in Bailey and Gillespie 2002.
3   Quoted in Sheahen 2004.
4   Quoted in "Is Science a Satisfying Replacement for Religion? A Conversation with E.O. Wilson" 2003.
5   Quoted in Bailey and Gillespie 2002. See also Pinker 2002, 189.
6   Quoted in Sheahen 2004.
7   Quoted in "Is Science a Satisfying Replacement for Religion? A Conversation with E.O. Wilson" 2003.
8   Pinker 2002, 189.
9   Dawkins 1976, 3.
10 Wilson 2002.
11 Wilson 1978, 4 and 6.
12 Quoted in Pinker 2002, 187.
13 Pinker 2002, 187-88.
14 Arnhart 1995 and 1998.
15 Lewis 1943, 16

1   Darwinian Conservatism?

1   Burke 1987, 79.
2   Dawkins 1996, 6.
3   Murray 2000, 47.
4   Arnhart 2000b, 26. See also Arnhart 2001.
5   Ibid.

6   Murray 2000, 47–48.
7   McGinnis 1997, 32; Arnhart 2000b, 27–28.
8   McGinnis 1997, 32; Arnahrt 2000b, 28.
9   Arnhart 2000b, 31.
10 McGinnis 1997, 32.
11 Murray 2000, 47–48; McGinnis 1997, 33–34.
12 McGinnis 1997, 31; Arnhart 2000b.
13 Salter 1996.
14 Wilson 2000, vi.
15 Arnhart 1995, 389.
16 Gelertner 1998, 52; Behe 2000, 28.
17 See Ferguson 2001; O'Sullivan 1999; Arnhart 2000b, 25.
18 See Bethell 2001; Ferguson 2001.
19 Tocqueville 2000, 7.
20 Ibid., 521.
21 Wilson 1993, vii.
22 Ibid., ix–xii.
23 Fukuyama 1999, 155–56.
24 Masters 1989, xiv. These claims of the public utility of the evolutionary account of morality are equally evident in the arguments of Larry Arnhart and Robert McShea, the other primary proponents of the new Darwinian political theory to whom this book gives considerable attention.
25 Tocqueville 2000, 417.
26 Wilson 1993, 220.
27 McShea 1990, 52.
28 Arnhart 1998, 249 (emphasis in original).
29 Tocqueville 2000, 417.
30 Arnhart 1998, 266.

2   *Decent Materialism*

1   Tocqueville 2000, 506.
2   Ibid., 506–07.
3   Ibid., 507.
4   Ibid., 510–11.
5   Ibid., 511–512.
6   Ibid., 515–516,
7   Ibid., 508–09.
8   Ibid., 240.
9   Ibid., 482–84.
10 Ibid., 236 and 409.
11 Ibid., 641.
12 Ibid., 517.
13 Ibid., 518–19.
14 Ibid., 419 and 280.
15 Ibid., 11.
16 Ibid., 264 and 280.
17 Ibid., 518 and 521.
18 Ibid., 281 and 519.

19 Ibid., 403.
20 Ibid., 47.
21 Ibid., 51.
22 Ibid., 404.
23 Ibid., 408.
24 Ibid., 454.
25 Ibid., 519.
26 Ibid., 409.
27 Ibid., 411 and 413.
28 Ibid., 426.
29 Masters 1989, 240.
30 McShea 1990, 7.
31 Masters 1989, 1.
32 Arnhart 1998, 67.
33 Ibid., 39.
34 Wilson 1993, 20.
35 Ibid., 22.
36 Arnhart 1998, 119. See also Fukuyama 2002, 142.
37 I hasten to add that it does not follow, and that I do not mean to suggest, that these unfavorable circumstances, by eliciting desires in conflict with the natural sense of obligation to preserve one's offspring, actually justify parents in killing their children
38 Arnhart 1998, 31-36.
39 Wilson 1993, xiii.
40 Arnhart 1998, 36-38. On the untenability of a strict nature-nurture dichotomy, see also Masters 1993, 119.
41 Ibid., 112.
42 Wilson 1993, 55-60.
43 Fukuyama 1999, 168.
44 Masters 1989, 155.
45 Fukuyama 1999, 162 and 172-73.
46 Ibid., 183-84. On this point see also Masters 1992, 84.
47 McShea 1990, 85.
48 Wilson 1993, 24.
49 Ibid., xv.

3 *Ennobling Democracy*

1 Fukuyama 1999, 6; emphasis in original.
2 Ibid. The scientific deficiencies of Hobbes and Locke are also discussed at some length by Larry Arnhart (1998) and Roger Masters (1989).
3 Ibid., 138 and 166-67.
4 Arnhart 1998, 4.
5 Wilson 1993, 13.
6 Masters 1989, xiv-xv.
7 Ibid., 149-50. Elsewhere, while praising the insights into the workings of the brain that are made possible by modern technology, Masters ridicules the notion that one could understand thought by thinking. Though made in the context of a criticism of contemporary psychology, this criticism would seem to apply no less to ancient philosophers like Plato and Aristotle, who had to rely to a considerable extent on reasoning from ordinary human experience.

8   Fukuyama 1999, 138. Similarly, in *Our Posthuman Future*, Fukuyama, while following Aristotle in seeking to establish principles of right on human nature, implies the superiority of the account of human nature provided by contemporary science. "Modern biology," he suggests, "is finally," after millennia of philosophic debate, "giving some meaningful empirical content to the concept of human nature" (Fukuyama 2002, 13).

9   Ibid., 273.

10 Wilson 1993, 240-42.

11 Ibid., 261 and 174.

12 Ibid., 137.

13 Fukuyama 1999, 253.

14 Ibid., 254.

15 Ibid., 253.

16 Montesquieu 1989, 338.

17 Quoted in Arnhart 1998, 75 and 76. It is true that in the latter passage quoted Darwin also speaks of the moral sense as "in later times" ruled by "deep religious feelings." Given his emphasis on social approval, however, Darwin seems to refer to religion as an institution guaranteeing social order rather than as a teacher of what is of absolute or transcendent obligation.

18 Wilson 1993, 121, 127, and 96.

19 *Politics* 1252a25-b27. All subsequent references to Aristotle's works similarly employ the Bekker numbers.

20 Ibid. 1252b29-30.

21 Ibid. 1281a2-4.

22 *Nicomachean Ethics* 1104b10-1105b22-24.

23 Ibid. 1115a10-15.

24 Ibid. 1108b23-26.

25 Ibid. 1109a12-19.

26 Ibid. 1128a13-16.

27 Arnhart 1995, 392. Here Arnhart and Fukuyama differ in that the former accepts the possibility of group selection while the latter rejects it (see Fukuyama 1999, 161). Hence Arnhart presents courage as arising from a natural inclination, while Fukuyama suggests that it somehow transcends the natural morality he discusses (see Fukuyama 1999, 235).

28 Arnhart 1998, 144.

29 Ibid., 137.

30 *Nicomachean Ethics* 1116a16-23.

31 Ibid. 1117b11-20.

32 Fukuyama 1999, 10-11.

33 Ibid., 275.

34 Hobbes 1991, 109.

35 Ibid., 111. Arnhart implicitly concedes the Hobbesian aspects of Darwinian natural right (1994, 474).

36 McShea 1990, 77.

37 Arnhart 1998, 70.

38 Ibid., 265.

39 We will consider more closely the Darwinian treatment of psychopathy, and its implications for the attempt to base morality only on the evolutionary account of our natural desires, in Chapter 6.

40 Masters 1989, 155. Masters similarly notes that whether organisms will cooperate "can usually be predicted by a cost-benefit calculus of the alternatives" (1989, 155).

41 Arnhart 1998, 160.
42 See, for example, Arnhart 1998, 220.
43 Masters 1989, 243.
44 Ibid., 232 (Fukuyama's emphasis). Elsewhere, however, Fukuyama acknowledges "folk religion" as a spontaneous development to support biologically grounded norms. See Fukuyama 1999, 152.
45 *Politics* 1328b10-15 (my emphasis). The public role of religion is also apparent in Aristotle's discussion of the virtue of magnificence in Book IV of the *Nicomachean Ethics.*
46 Fukuyama 1999, 231-32.
47 *Nicomachean Ethics* 1099a 6-21.
48 Nicomachean Ethics 1179a30-32.
49 *Politics* 1252b31-34.
50 See *Nicomachean Ethics* 1177a12-1178a2.
51 Tocqueville 2000, 431.
52 Ibid., 437and 433..
53 Ibid., 502.
54 Ibid., 283-84. Elsewhere he speaks of the "instinct for another life" that "delivers" men "to the foot of altars" and their "hearts to the precepts and consolations of faith" (Ibid., 287).
55 Fukuyama 1999, 278 and 279.
56 Arnhart 1998, 273.
57 See the *Nicomachean Ethics* 1098a1-17.
58 Wilson 1993, 33.
59 Ibid., 30.
60 Ibid., 44 (Wilson's emphasis).
61 Ibid., 30.
62 Masters 1999, 230.
63 Ibid., 225.
64 Ibid., 152 and 230.
65 Ibid., 19-20.
66 Ibid., 230.
67 McShea 1990, 17.
68 Ibid., 13.
69 Ibid., 211.
70 Consider Lewis 1947, 46-47, note 2. The objections to McShea's theory offered in this paragraph are essentially the same as those posed by Lewis in response to all efforts to erect a theory of morality on the basis of the satisfaction of feelings.
71 Arnhart 1998, 270.
72 Ibid., 275.
73 Fukuyama 1999, 235.
74 Ibid., 232.

4 *Moral Universalism*

1 Arnhart 1998, 77. On this point see also Ridley's *The Origins of Virtue*, 69-70.
2 Fukuyama 1999, 236.
3 Quoted in Arnhart 1998, 144.
4 Wilson 1993, 192.
5 Arnhart 1998, 143.
6 Fukuyama 1999, 213-16.

7   Ibid., 233.

8   Ibid., 236 and 234.

9   Ibid., 234.

10   Wilson 1993, 54.

11   McShea 1990, 179.

12   Masters 1989, 190.

13   Fukuyama 1999, 280.

14   Ibid., 243.

15   Ibid.

16   Ibid., 237.

17   Ibid.

18   Ibid., 240.

19   Ibid., 279 and 280-81.

20   Ibid., 281.

21   Fukuyama makes a similar argument in his most recent book. In *Our Posthuman Future* (2002, 125-127) he contends that the logic of human nature points in the direction of the continued development of cooperative globalization.

22   McShea expressly repudiates universalism (1990, 249), but, as the passages quoted in this paragraph illustrate, he has clear aspirations in this direction.

23   McShea 1990, 255 and 148.

24   Ibid., 255.

25   Ibid., 263.

26   Ibid., 78.

27   Ibid., 247.

28   Ibid., 175.

29   Ibid., 194-95. Similarly, Matt Ridley indicates that humans, like chimps, tend to be xenophobic and that tribal conflict is therefore inevitable. See *Origins of Virtue* 165-66.

30   Alexander *Darwinism and Human Affairs*, xiii.

31   Wilson 1993, 54.

32   Ibid., 50 and 191-93.

33   Ibid., 211. For the whole argument see 191-212.

34   Ibid., 44 and 128.

35   Ibid., 215.

36   Ibid., 226.

37   Ibid., 228.

38   Masters 1989, 231.

39   Ibid., 228.

40   Ibid., 225.

41   Ibid., 190.

42   Arnhart 1998, 146-47.

43   Ibid., 148.

44   Ibid., 34. E.O. Wilson similarly speaks of the "biological joy of warfare" in *Sociobiology*, 573.

45   Ibid., 149.

46   Ibid., 62.

47   Masters 1989, 151 and 18-19.

48   Ibid., 225.

49   Ibid., 232.

50   Ibid., 186.
51   Ibid., 222–23.
52   Ibid., 245.
53   Ibid., 21.
54   Ibid., 190.
55   Ibid., 226.
56   McShea 1990, 242–43.
57   Ibid., 245.
58   Ibid., 245.
59   Arnhart 1998, 172.
60   Ibid., 162 and 231.
61   Ibid., 160.
62   Ibid., 167–68.
63   Ibid., 173.
64   Ibid., 207–08.
65   Ibid., 209. Similarly, E.O. Wilson notes that human flexibility is such that even very dysfunctional societies, such as the Jamaican slave system, can persist for a long time. *Sociobiology* 549.
66   Ibid., 207–08.
67   Ibid., 34 and 148.
68   Ibid., 162 and 169.
69   Ibid., 171.
70   Ibid., 170 and 176.
71   Wilson 1993, 52–53.
72   Wright 1994, 73–74.
73   Ibid., 265. See also 132–137.
74   Hobbes 1991, 88. On the Darwinian modification of Hobbes's understanding of the state of nature, see Fukuyama 1999, 165.

5   *Darwinism and the Family*

1   Fukuyama 1999, 243.
2   Tocqueville 2000, 535 and 538.
3   Fukuyama 1999, 244.
4   Ibid., 16.
5   Ibid., 36–37.
6   Ibid.
7   Wilson 1993, 220.
8   Ibid., 146–47.
9   Ibid., 162–63.
10   Arnhart 1998, 52–53.
11   Ibid., 75.
12   Ibid., 56.
13   Fukuyama details these familial and social trends in Chapter 2 of *The Great Disruption*.
14   Fukuyama 1999, 45.
15   Wilson 1993, 159.
16   On these issues see Fukuyama 1999, 93.
17   On these issues see Fukuyama 1999, 101–111.

18   Fukuyama 1999, 13.
19   Arnhart 1998, 97.
20   Ibid., 97 and 98.
21   Ibid., 98.
22   Ibid., 92-93.
23   Ibid., 93-94.
24   Ibid., 95.
25   Ibid., 103.
26   Wilson 1993, 158. See also 126.
27   Arnhart 1998, 113.
28   As Arnhart rightly points out, the evolutionary account does not suggest that parenting is experienced as unproblematically pleasant by human beings. Its obligations clearly conflict with many other natural desires: the diligent care of children diminishes, for example, the amount of time one can dedicate to the pursuit of social status through professional or political achievement, to say nothing of its impositions on the desire for ordinary ease and comfort. As a natural desire, however, parental care compensates to some extent for the pains it requires, and it cannot be abandoned without the experience of a serious sense of frustration and loss. See Arnhart 1998, 89 and 113.
29   Arnhart 1998, 121.
30   Tocqueville 2000, 279.
31   Ibid., 573.
32   Ibid., 573-74.
33   Ibid., 574.
34   Ibid., 565-66.
35   Ibid., 574.
36   Ibid.
37   Arnhart 1998, 123. Wilson (1993, 165) similarly suggests that men are by nature more aggressive than women.
38   Arnhart 1998, 99.
39   Wilson 1993, 183-85.
40   Arnhart 1998, 129-30.
41   On these issues, see Arnhart 1998, 131-132.
42   Tocqueville 2000, 576.
43   Arnhart 1998, 143. Wilson 1993, 179.
44   Tocqueville 2000, 575.
45   Ibid., 278-79. See also 567-73.
46   Arnhart 1998, 134.
47   Ibid., 265.
48   Ibid., 134.
49   For a discussion of the different mating desires of men and women, see Arnhart 1998, 132-37.
50   Fukuyama 1999, 98 (footnote).
51   Arnhart 1998, 265.
52   Fukuyama 1999, 97.
53   Ibid.
54   Ibid., 98 (footnote).
55   Ibid., 179.
56   See Fukuyama 1999, 95-96 and 100-101, and Wilson 1993, 176.
57   Arnhart 1998, 123 and 136.

58 Ibid., 136-37.
59 Ibid., 136 (my emphasis).
60 Such an understanding of marital commitment is still compatible with democracy, however, because, as Tocqueville points out, no one is compelled to enter into marriage and everyone is allowed to choose his or her partner freely.
61 Fukuyama 1999, 121.
62 Tocqueville 2000, 2798-79.
63 Ibid., 571.
64 Ibid., 279.
65 Wilson 1993, 163.
66 See Wilson 1993, 169-70.
67 Ibid., 165.
68 Ibid., 176.
69 Fukuyama 1999, 99.
70 Arnhart 1998, 135-36.
71 In Volume II of *Democracy in America* Tocqueville also contends that democracy itself tends to generate an intolerance of adultery. Because everyone is free to choose whether and whom to marry, Tocqueville notes, public opinion has little sympathy for those who violate obligations that they took on voluntarily. Nevertheless, as our recent social history, outlined by Fukuyama in *The Great Disruption*, indicates, equality of conditions alone does not seem sufficient to sustain the strict understanding of marriage that Tocqueville defends.
72 Tocqueville 2000, 278-79.

6 *The Demise of Darwinian Morality*

1 McShea 1990, 15 and 17.
2 Arnhart 1998, 17.
3 McShea 1990, 206-07.
4 Arnhart 1998, 70.
5 Roger Masters (1989, 240) offers a similar naturalistic response to positivism, as does Francis Fukuyama (2002, 114-117).
6 Ibid., 74.
7 McShea 1990, 49-63.
8 Ibid., 35.
9 Wilson 1993, 238-39.
10 Lewis 1947, 15-16.
11 Arnhart 1998, 70 (Arnhart's emphasis).
12 Ibid. (Arnhart's emphasis). Here Arnhart relies on Nicholas Capaldi's interpretation of Hume's moral theory. See Capaldi 1989.
13 McShea 1990, 220.
14 Ibid., 30 and 213.
15 Arnhart 1998, 70.
16 Ibid., 211.
17 McShea 1990, 216-17. Arnhart also recognizes these two causes of feeling deviance, genetic abnormality and unusual environmental conditions, and discusses their respective manifestations in "primary" and "secondary" psychopathy (1998, 220-221).
18 Wilson 1993, 107.
19 Arnhart 1998, 229.
20 McShea 1990, 227.

21 Ibid., 76. See also page 15, where McShea asserts that, feelings being the only ground of value judgments, there is "no basis on which we can mount a moral criticism of someone who pursues" the satisfaction of his feelings "effectively and with the most singleminded intensity."
22 Wilson 1993, vii and ix.
23 Arnhart 1998, 212.
24 McShea 1990, 227.
25 Ibid., 4-5.
26 Ibid., 107.
27 Arnhart 1998, 216.
28 Ibid., 222-23.
29 Ibid., 222.
30 Ibid., 222.
31 Ibid., 221 and 223.
32 Ibid., 220.
33 Ibid., 221 and 223.
34 Ibid., 82.
35 McShea 1990, 218.
36 Fukuyama 1999, 176.
37 Masters 1989, 31.
38 Ibid., 9 and 179.
39 Ibid., 164.
40 Arnhart 1989, 147-48.
41 Wilson 1993, 11 and 238.
42 Arnhart 1998, 75. See also 82 and 136.
43 McShea 1990, 260-61.
44 Ibid., 255.

7 *The Abolition of Man*

1 Lewis 1947, 69.
2 See Kass and Wilson 1998, 40-42.
3 Tocqueville 2000, 514.
4 Nietzsche 1989, 116.
5 Kass and Wilson 1998, 19.
6 Fukuyama 1999, 146.
7 Arnhart 1998, 237.
8 Masters 1989, 119 and 70.
9 Wilson 1993, 41 (my emphasis).
10 McShea 1990, 175.
11 Arnhart 1998, 245. On this distinction see also Arnhart 1999. Masters similarly suggests that "natural purpose is immanent in living things, not extrinsic to them" (1989, 248).
12 Ibid., 243.
13 McShea 1990, 74.
14 Masters 1989, 246.
15 Quoted in Arnhart 1998, 180.
16 Arnhart 1998, 181.

17 Arnhart regards such natural slavery as merely hypothetical for Aristotle as well. While Aristotle seems explicitly to confirm the existence of natural slaves, Arnhart contends that his argument is actually intended to bring to light the incoherence of any defense of human slavery. See Arnhart 1998, 172.

18 Arnhart 1998, 275.

19 McShea 1990, 191.

20 On this point see Ridley 1996, 19.

21 Ibid., 84-85.

22 Ibid., 195.

23 Ibid., 197.

24 Ibid., 198.

25 Ibid., 212.

26 Ibid., 233.

27 Ibid., 234.

28 Fukuyama 2002, 77-78.

29 Ibid., 98. See also McShea (1990, 191), who contends that it is not "practicable to select particular disfavored passions for excision, for although each passion operates in us independently, the passions are genetically intertwined. Our capacity to feel anger, for instance, seems inextricably connected to our capacity to feel affection."

30 Fukuyama 2002, 56.

31 Masters 1989, 128.

32 McShea 1990, 189.

33 Ibid., 203.

34 Ibid., 204.

35 Masters 1989, 237.

36 It is worth noting, in passing, that Hobbes also holds, like many modern scientists, that free will is merely an illusion created by the multiplicity of factors that determine human behavior.

37 Masters 1989, xv.

38 See *The Ethics of Human Cloning* (Kass and Wilson, 1998). Wilson's only concerns about cloning are that human clones might be used for bad purposes and that excessive cloning might diminish that natural genetic variation that is necessary for the continued flourishing of the species. He suggests that we might prevent such bad consequences of cloning by insisting that all clones be born to a married woman and that people only be allowed to clone from their own family. Such concerns are utilitarian, and Wilson seems indifferent to Kass's contention that cloning is, as a radical rejection of natural reproduction and alteration of the parent-child relationship, morally problematic in itself.

39 Fukuyama 2002, 100.

40 Ibid., 88.

41 Ibid., 91.

42 Quoted in Fukuyama 2002, 161.

43 Fukuyama 2002, 170, 160-61.

44 Ibid., 168. For the whole of Fukuyama's argument about the irreducible complexity of natural systems, see pp. 160-171.

45 Ibid., 171.

46 Ibid., 172.

47 Fukuyama 1999, 146.

48 Fukuyama 2002, 152 and 162. While Fukuyama argues against those who contend that such Darwinian principles justify the deliberate modification of human nature and who contend that such principles are sufficient to explain the complexity of human nature, he appears to side with them in embracing such principles as true.

49 Indeed, insofar as his argument shows a variety of natural non-human systems whose complexity cannot be explained on the basis of a reductionistic account of the operation of their parts, it is not clear that Fukuyama succeeds even in demonstrating human uniqueness.

50 Fukuyama 2002, 207.

51 Ibid., 208-09. Here Fukuyama is following Leon Kass's argument in *Toward a More Natural Science* (1985).

52 Ibid., 156-57.

53 See Fukuyama 2002, 150   and 155.

54 Fukuyama 2002, 156.

55 Ibid., 173 and 218.

56 Ibid., 14 and 106.

57 Fukuyama 1999, 281.

58 Ibid., 10-11.

8 Beyond Darwinism

1 See Book II of *The Republic*.

2 McGinnis 1997, 33, 31-32.

3 Ibid., 32.

4 McGinnis 1999, 32. See also 1997, 33-34.

5 See McGinnis 1997.

6 McGinnis is aware of this limitation. In response to David Gelertner's friendly criticism he states that "our moral objectives" cannot "be directly derived from science, and thus they cannot be changed by science" (1997, 54). In this he departs from Arnhart and McShea, who, among the Darwinian political theorists, most clearly contend that our scientific knowledge of human nature is a sufficient basis for a theory of right.

7 McGinnis 1997, 34-35.

8 Arnhart 2000, 25 and 27. Arnhart makes essentially the same criticism of Richard Hassing. Because Hassing (1999-2000) contends that cosmic teleology appears both necessary to morality but also intellectually difficult to defend, Arnhart classes him with Straussians who offer cosmic teleology as a "noble lie" necessary to a wholesome politics (Arnhart 2000a).

9 On this point see McGinnis 1997, 54.

10 Ferguson 2001.

Works Cited

Alexander, Richard. 1979. *Darwinism and Human Affairs*. Seattle: University of Washington Press.

Aristotle. 1990. *The Nicomachean Ethics*. Translated by H. Rackham. Cambridge, Massachusetts: Harvard University Press.

_____. 1984. *The Politics*. Translated by Carnes Lord. Chicago: The University of Chicago Press.

Arnhart, Larry. 2001. "Stealing Darwin." *National Review*, 2 April.

_____. 2000a. "Defending Dariwnian Natural Right." *Interpretation* 27, No. 3: 263-277.

_____. 2000b. "Conservatives, Darwin, and Design: An Exchange." *First Things*, November: 23-28.

_____. 1999. "Roger Masters: Natural Right and Biology." In *Leo Strauss, the Straussians, and the American Regime*, edited by Kenneth L. Deutsch and John A. Murley. Lanham, Maryland: Rowman and Littlefield, 293-303.

_____. 1998. *Darwinian Natural Right: The Biological Ethics of Human Nature*. Albany, New York: State University of New York Press.

_____. 1995. "The New Darwinian Naturalism in Political Theory." *American Political Science Review* 89: 389-400.

_____. 1993. "The Darwinian Biology of Aristotle's Political Animals." *American Journal of Political Science* 38, No. 2: 464-85.

Bailey, Ronald. 1997. "Origin of the Specious: Why do Neoconservatives Doubt Darwin?" *Reason*, July: 22.

Bailey, Ronald and Nick Gillespie. 2002. "Biology vs. the Blank Slate: Evolutionary Psychologist Steven Pinker Deconstructs the Great Myths about how the Mind Works." *Reason*, Volume 34, Issue 5.

Bethell, Tom. 2001. "Against Sociobiology." *First Things*, January: 18-24.

Burke, Edmund. 1987. *Reflections on the Revolution in France*, edited by J.G.A. Pocock. Indianapolis: Hackett Publishing Company.

Capaldi, Nicholas. 1989. *Hume's Place in Moral Philosophy*. New York: Peter Lang.

Dawkins, Richard. 1996. *The Blind Watchmaker*. New York: W.W. Norton and Company.

————. 1989. *The Selfish Gene*. New York: Oxford University Press.

Ferguson, Andrew. 2001. "Evolutionary Psychology and Its True Believers." *The Weekly Standard*, 19 March.

Fukuyama, Francis. 2002. *Our Posthuman Future: Consequences of the Biotechnology Revolution*. New York: Farrar, Straus, and Giroux.

————. 1999. *The Great Disruption: Human Nature and The Reconstitution of Social Order*. New York: The Free Press.

Gelertner, David. 1998. "The Decsent of Man: Can Conservative Political Concepts be Derived from Evolution? Critics Respond to John O. McGinnis." *National Review*, 9 March.

Hassing, Richard F. 1999. "Darwinian Natural Right?" *Interpretation* 27, No. 2: 129-160.

Hobbes, Thomas. 1991. *Leviathan*, edited by Richard Tuck. New York: Cambridge University Press.

"Is Science a Satisfying Replacement for Religion? A Conversation with E.O. Wilson." 2003. *Research News and Opportunities in Science and Theology*, July/August issue. Posted on *beliefnet*: www.beliefnet.com/story/131/story_13113.html

Kass, Leon and James Q. Wilson. 1998. *The Ethics of Human Cloning*. Washington, D.C.: The American Enterprise Institute Press.

Kass, Leon. 1985. *Toward a More Natural Science*. New York: The Free Press.

Lewis, C.S. 1947. *The Abolition of Man: How Education Develops Man's Sense of Morality*. New York: MacMillan Publishing Company.

————. 1943. *Mere Christianity*. New York: Simon and Schuster.

Masters, Roger D. 1993. *Beyond Relativism: Science and Human Values*. Hanover and London: University Press of New England.

————. 1992. "Naturalistic Approaches to Justice in Political Philosophy and the Life Sciences." In *The Sense of Justice: Biological Foundations of Law*, edited by Roger D. Masters and Margaret Gruter. London: Sage Publications.

_____. 1989. *The Nature of Politics*. New Haven: Yale University Press.

McGinnis, John O. 1999. "Unnatural Selection: The Feminists' Unconvincing Biology." *National Review*, 19 April, 30-32.

_____. O. 1997. "The Origin of Conservatism." *National Review*, 22 December, 31-35.

McShea, Robert. 1990. *Morality and Human Nature: A New Route to Ethical Theory*. Philadelphia: Temple University Press.

Montestquieu. 1989. *The Spirit of the Laws*. Translated by Anne M. Cohler, Basia Carolyn Miller, and Harold Samuel Stone. New York: Cambridge University Press.

Murray, Charles. 2000. "Deeper into the Brain." *NationalReview*, 24 January, 46-49.

Nietzsche, Friedrich. 1989. *Beyond Good and Evil*. Translated by Walter Kaufmann. New York, Vintage Books.

O'Sullivan, John. 1999. "Types of Right: How the Conservatives Break Down." *National Review*, 11 October, 20-21.

Plato. 1968. *The Republic of Plato*. Translated by Allan Bloom. New York: Basic Books.

Pinker, Steven. 2002. *The Blank Slate*. New York: Viking.

Ridley, Matt. 1999. *Genome: The Autobiography of a Species in 23 Chapters*. New York: Perennial.

_____. 1996. *The Origins of Virtue: Human Instincts and the Evolution of Cooperation*. New York: Penguin Books.

Salter, Frank. 1996. "Political Science: Sociology's Disdain for the Scientific Method Undermines its Scientific Pretensions." *National Review*, 3 June, 45.

Sheahen, Laura. 2004. "Religion: For Dummies: Scientist Richard Dawkins on Darwin, the Sistine Chapel, and Why the World Would be Better Off Without Religion." An online interview on *beliefnet*: www.beliefnet.com/story/136/story_13688.html

Tocqueville, Alexis de. 2000. *Democracy in America*, translated by Harvey C. Mansfield and Delba Winthrop. Chicago: University of Chicago Press.

Wilson, E.O. 2002. *The Future of Life*. New York: Alfred A. Knopf.

_____. 2000. *Sociobiology: The New Synthesis*. Cambridge, Massachusetts: Harvard University Press.

_____. 1978. *On Human Nature*. Cambridge, Massachusetts: Harvard University Press.

Wilson, James Q. 1993. *The Moral Sense*. New York: The Free Press.

Index

The Abolition of Man (C.S. Lewis), 135, 149
abortion, 78
Achilles, 65
Adeimantus, 178
adultery, 118, 124, 126-27, 132, 200
aggression, 116, 144, 165, 199
Alexander, Richard, 79, 197
altruism, 37, 38, 45, 63, 68, 74, 86
America, 5, 15, 18-20, 24, 26-29, 82, 83, 98, 100-02, 104-06, 111-14, 117-18, 121, 126, 128, 131-32, 152
ancient political philosophy, 42-44, 194
aristocracy, 10, 14, 18-19, 22-23, 25, 27-29, 101, 125, 148, 152, 173
Aristotle, 6, 42-53, 56-62, 66, 77, 86, 103, 150-51, 158, 194-95, 202

Arnhart, Larry, 6-8, 11, 13, 34-36, 43, 47, 50, 53-56, 60-61, 67-68, 72-73, 83-85, 86, 90-95, 103-04, 107-09, 110, 114-16, 118-19, 121, 123-24, 130, 133-35, 137-38, 140-43, 145-46, 155-56, 158-59, 161-62, 180, 183, 186-89, 191-204
atheism, 16, 26
Augustine, Saint, 144

Bach, J.S., 141
Behe, Michael, 191
Bethell, Tom, 191
biotechnology, 148, 150-54, 158-61, 168-69, 171-76
Brave New World (Aldous Huxley), 175
Burke, Edmund, 3, 5, 45-46, 186, 191
Calvin, John, 144
Capaldi, Nicholas, 201

capitalism, 46-47, 175
Catholic Church, 54, 80, 121, 125
Chesterton, G.K., 128
children, 27, 34, 36, 52, 56, 57, 79, 102-23, 127-32, 138-39, 149-51, 171-72, 174, 179, 181-82, 193, 199, 203
Christianity, 10, 26, 74, 80, 84, 131-32, 169, 187
Clinton, William, 100
cloning, 149-50, 154, 168, 172, 203
competition, 4, 6-7, 19, 38, 50, 73, 89, 94, 118, 142, 144, 162, 165, 174
conservatism, 3-10, 13-15, 26-27, 100-01, 104, 114, 154, 178-89
constitutionalism, 86-90
cooperation, 37-39, 42, 45, 50, 55-57, 64-65, 72-73, 79, 86, 91, 94-96, 101-03,

110, 112, 116, 131, 143-44, 163, 165, 188, 195, 196

crime, 39, 52, 104, 118, 140, 146

cultural relativism, 12, 32-37, 146

Darwin, Charles, 4, 43, 47, 48, 50, 67, 72, 84, 103, 109, 189, 194

Darwinian Natural Right (Larry Arnhart), 13, 90, 161

Darwinism and Human Affairs (Richard Alexander), 197

Dawkins, Richard, 4, 191

democratic materialism, 10, 15, 18-21, 25, 26, 31, 40, 41, 42, 43, 53, 70, 71, 77, 99, 152-54, 181

democracy, 9, 10, 14, 15, 17-29, 41, 56, 59, 85, 86, 95, 100-01, 112, 117, 125, 127, 133, 148, 152-54, 175-77, 186, 200

Democracy in America (Alexis de Tocqueville), 10, 59, 111, 112, 118, 126, 200

Descartes, Rene, 27

despotism, 17, 20, 148, 152, 157-61

divorce, 54, 95, 104, 118-19, 182

The End of History and the Last Man (Francis Fukuyama), 76

The Enlightenment, 45-46, 75

equality, 15, 22-25, 27, 30, 82, 97, 112-13, 116-18, 148, 152-54, 173-74, 200

The Ethics of Human Cloning (Leon Kass and James Q. Wilson), 203

Europe, 26, 80, 100, 104, 106, 111-13, 118, 126-28

evolution, 7, 16, 18, 30, 38, 44, 63, 68, 76, 83, 99, 109, 110, 119, 144, 155-57, 170

evolutionary psychology, 6, 31

exploitation, 74, 82, 85, 91-92, 93, 94-96, 98, 142, 158-60, 163, 180, 182

family, 7-8, 22, 47, 52, 65, 67, 71, 80, 99-132, 146, 179, 181-82, 203
  as rooted in nature, 34-35, 43, 101, 102, 106, 107-12, 179, 181-82
  as school of virtue and sociability, 13, 102-04, 111-14

Ferguson, Andrew, 186, 188, 191, 204

First Things, 6

freedom, 9, 10, 17, 20, 24, 26, 70, 95, 100, 105, 148-50, 154, 157-58, 160, 181

free will, 9, 166-67, 169, 186, 203

Freud, Sigmund, 86

Fukuyama, Francis, 8, 11, 12, 37, 38, 42, 44-47, 52, 57, 60, 61, 68-69, 72-73, 75-77, 78, 100, 101-03, 105, 106, 120-23, 126, 129-30, 144, 149, 154, 155, 165, 168-77, 192-204

fundamentalism, 16

Gelertner, David, 191, 204

genetic fitness, 37, 39, 55-57, 144

genocide, 73, 82

Germans, 73

Glaucon, 131, 178

God, 4-6, 9, 12-13, 21, 24, 57, 132, 134, 155-56, 169-70, 177

The Great Disruption, (Francis Fukuyama), 12, 45, 52, 77, 102, 121-22, 169, 171, 173, 175, 186, 198, 200

Hare, Robert, 141-42

Hassing, Richard, 204

Hector, 65-67

hedonism, 17-18, 21, 26, 31, 32, 41, 43, 62, 100, 101, 181

Heretics (G.K. Chesterton), 128

hierarchy, 22, 88, 120, 156, 170

Hobbes, Thomas, 42, 48, 52-53, 82, 96, 167-68, 193, 195, 198, 203

Homer, 66

human dignity, 9, 10, 17, 24, 73, 75-76, 148, 150-54, 160, 169, 171, 173-74, 181

human nature, 30, 32, 41, 44, 49, 55, 57, 69, 73, 74, 76, 78, 81, 83, 85, 87, 90, 96, 107, 139, 144, 184-87, 194, 196, 203, 204
  as including a longing for the transcendent, 19, 21, 58-61, 62
  biotechnological manipulation of, 149-77
  Darwinian account of, 9, 10, 70, 71-72, 80, 82, 84, 86, 94, 95, 98, 99, 111, 117, 119-22, 124, 133, 154, 155, 157, 163, 173, 175, 184, 187
  as supporting conservatism, 6, 7, 8, 101, 178-81
  as inherently sociable and moral, 7, 32, 37, 39, 42, 43, 44, 47, 146

Hume, David, 134-35, 158-59, 201

The Iliad (Homer), 65-66

illegitimacy, 104

individualism, 22-23, 71, 80
infanticide, 34-35
Islam, 169
Israel, 107, 115

John Paul II, Pope, 170, 177
Judaism, 169
justice, 7, 10, 22, 24-25, 31, 33,
35, 37, 40, 43, 45, 63-64,
71, 74, 78, 82, 84-88, 90,
93-95, 98, 131, 148, 153,
158-59, 163, 167, 179, 183-
85, 188

Kant, Immanuel, 45, 134, 144
Kass, Leon, 150, 154-55,
202-04
kibbutzim, 107-08, 115
kin, 9, 30, 37, 55, 56, 63, 64,
74, 80, 83, 85, 88, 89, 93,
95, 102, 111, 158, 182

Lewis, C.S., 67, 135, 149, 161,
196, 200, 202
liberalism, 8, 15, 26, 76-77,
104
Locke, John, 42, 82, 193

McGinnis, John, 179-84,
191, 204
Machiavelli, Niccolo, 141
Machiavellian personality,
141-45, 147, 183
McShea, Robert, 11, 13, 32,
39, 53-54, 65-67, 74, 77-79,
88-90, 134-36, 137-40,
142, 145-46, 156-57, 161-
64, 166-67, 183, 192-93,
195-97, 200-03
Madison, James, 180
majority tyranny, 10, 15,
17, 22-24, 25-26, 31, 40,
70, 71, 85-86, 90, 95-98,
100-01, 181
marriage, 12, 80, 95, 104-06,
118-124, 200

Masters, Roger, 11, 12, 32, 33,
38, 43, 44, 55, 56, 63-65,
69, 74, 82-83, 86-88, 144,
155, 157, 166-68, 192-93,
195-97, 200, 202-03
Mealy, Linda, 141
The Metaphysics (Aristotle), 58
modern political philoso-
phy, 42-43, 52, 82
Montesquieu, 46, 194
The Moral Animal (Robert
Wright), 94
moral feelings, 30, 34, 38-39,
47-48, 50, 53-54, 61, 63,
65-67, 78-79, 83, 85,
89-90, 93, 94, 95, 96, 98,
133-47, 163
moral relativism, 12, 31-32,
40, 41, 99, 133, 134-36,
139, 178
moral sense, 35, 36, 38, 39,
43, 47, 72, 80, 81, 90, 137-
38, 155, 189
The Moral Sense (James Q.
Wilson), 11, 39, 145
morality, 4, 8, 11, 12, 13, 30,
31, 32, 35, 37, 44, 52, 64,
72, 74, 84, 86, 89, 90, 104,
133-47, 174, 183, 184-87,
189, 195, 196
as artificial, 33-37, 42, 43,
146, 167, 178
as decency or sociability,
41, 44, 45, 47, 53, 57, 64,
67, 70, 101, 181, 188
as nobility or excellence,
41, 42, 43, 44, 47-53, 59,
62, 99-100, 176, 185, 188
as rooted in human na-
ture, 8, 12, 33-37, 52, 53, 57,
69-70, 73, 96, 146, 178
universal, 71-98, 99-101
*Morality and Human Na-
ture* (Robert McShea),
13, 162

Mother Theresa, 188
Murray, Charles, 6, 7, 191

National Review, 6, 179-80,
182-84, 191, 204,
natural selection, 4, 7, 16,
30, 32, 39, 55, 63, 72, 80,
81, 82, 109, 130, 142, 144,
155, 157,
171, 187
nature
as perfection (Aristotle),
53, 56-58
relationship to nurture,
35-36
universality of, 33, 53-54
variability of, 33-35
The Nature of Politics
(Roger Masters), 12, 168
Nazis, 73, 74
nepotism, 74, 93, 94, 102
The Nicomachean Ethics
(Aristotle), 48-51, 58, 66,
194-95
Nietzsche, Friedrich, 32, 154,
186, 202
Noyes, John Humphrey, 108

Oneida Community, 108-
09, 115
The Origins of Virtue (Matt
Ridley), 196, 197
O'Sullivan, John, 191
Our Post-Human Future
(Francis Fukuyama),
76, 149, 168, 169, 171, 175,
194, 196

parental care, 103-04, 107,
108, 110, 119, 123-24, 179,
181, 199
philosophy, 67-68
Plato, 46-47, 131, 156, 178,
194
The Politics (Aristotle), 47,
194-95

populism, 5, 14-15
property, 8, 27, 36, 47, 69, 92, 179-80, 182
psychopathy, 54-55, 136-41, 147, 178, 183, 188, 195, 201

racism, 74, 78
reason, 4, 27-28, 52, 90, 91, 96, 134-35, 157, 158, 171
reciprocity, 8, 9, 31, 35, 37, 38, 39, 45, 55, 56, 63, 64, 71, 91, 93, 94, 95, 98, 101, 103, 146, 158-59, 163, 180, 182-83
religion, 40, 56-61, 68, 71, 75, 97, 99, 169, 173, 176, 185, 186-89, 194, 195
  as necessary to a decent democracy, 3, 4, 12, 13, 17, 24-27, 29, 31, 32, 41, 43, 70, 98, 177
  as occasionally fanatical in democracy, 19
  "of pity" (Nietzsche), 154
  as source of universal morality, 76
  as supporting family stability, 131-32
  as undermined by Darwinism, 3-4, 7
  as undermined by democratic social conditions, 27-29, 31
reproduction, 30, 37, 45, 47, 53, 55, 57, 61, 63, 102, 109, 110, 116-17, 119, 121, 129-30, 150, 155, 162, 180, 187-88, 203
The Republic (Plato), 131, 178, 204

Ricard, Samuel, 46
Ridley, Matt, 196, 197, 202

Salter, Frank, 191
self-interest, 6, 22, 31, 32, 37, 38, 40, 45, 52, 53, 55-56, 59, 89, 94, 97-98, 99, 125, 176, 180, 182
sex, 39, 52, 103, 106, 118, 120, 121-24, 126-27, 129-32, 154, 182, 184,
sex differences, 7-8, 112-118, 179, 199
slavery, 74, 83, 88, 90-95, 158-61, 198, 202
sociability, 36, 45, 47, 48-49, 57, 62, 74, 90, 96, 104, 107, 132, 175, See also morality
  as artificial, 42, 69-70
  as natural, 42, 44, 51, 72, 86, 87, 96, 101, 102, 178
social Darwinism, 4-5, 7
sociobiology, 6, 8, 31, 144
Sociobiology: The New Synthesis (E.O. Wilson), 8, 197, 198
Socrates, 178-79
Soviet Union, 73
Spirit of the Laws (Montesquieu), 46
Straussians, 204
survival, 9, 30, 37, 39, 45, 48, 53, 57, 67, 82, 109-10, 156, 162
sympathy, 7, 30, 35, 39, 47, 63, 71, 74, 80, 81, 93, 94, 96, 98, 100, 103, 115, 143, 146, 183, 188

teleology
  cosmic, 154, 156-57, 161,

165, 166, 171-73, 176-77, 184-87, 204
immanent, 156, 202
Thatcher, Margaret, 115
Tocqueville, Alexis de, 9-10, 12, 13-15, 17-31, 32, 41, 43, 53, 56, 59-62, 64, 70, 71, 74, 77, 85, 86, 95, 97-98, 99-102, 111-14, 117, 118, 121, 124-28, 131-32, 148, 152-, 175, 181, 185-86, 191-92, 195, 198-200, 202
totalitarianism, 73, 87-88
Toward a More Natural Science (Leon Kass), 204
Trivers, Robert, 144

United States, 26, 113, 118
utilitarianism, 110, 169, 171, 174, 203

virtue, 12, 24, 42, 43, 45, 47-53, 55, 57, 58, 62, 63, 64, 65, 68-69, 77, 102, 150, 175-76

Waal, Frans de, 144
war, 42, 72, 74, 79, 84, 92, 96
Weaver, Richard, 186
The Weekly Standard, 186
Wilson, E.O., 8, 191, 197, 198
Wilson, James Q., 8, 11-13, 34-36, 38, 39, 43, 44, 47, 62-63, 68, 72, 74, 80-81, 94, 103, 105, 109-10, 115-16, 118, 123, 128-30, 134-35, 137-38, 145, 155, 168, 192-96, 198-203
Wright, Robert, 94

This book was designed and set into type
by Mitchell S. Muncy,
with cover design by Stephen J. Ott,
and printed and bound
by Thomson-Shore, Inc.,
Dexter, Michigan.

❦

The cover illustration of Charles Darwin
is courtesy of North Wind Picture Archives.

❦

The text face is Adobe Caslon,
designed by Carol Twombly,
based on faces cut by William Caslon, London, in the 1730s,
and issued in digital form by Adobe Systems,
Mountain View, California, in 1989.

❦

The paper is acid-free and is of archival quality.

46